COMBUSTION AERODYNAMICS

FUEL AND ENERGY SCIENCE SERIES

edited by

J. M. BEÉR

Professor of Chemical Engineering and Fuel Technology
University of Sheffield (Great Britain)

COMBUSTION AERODYNAMICS

J. M. BEÉR

Professor of Chemical Engineering,
MASSACHUSETTS INSTITUTE OF TECHNOLOGY

and

N. A. CHIGIER

Benedum Professor, Department of Mechanical Engineering,
CARNEGIE-MELLON UNIVERSITY

ROBERT E. KRIEGER PUBLISHING COMPANY
MALABAR, FLORIDA
1983

Original Edition 1972
Reprint Edition 1983

Printed and Published by
ROBERT E. KRIEGER PUBLISHING COMPANY, INC.
KRIEGER DRIVE
MALABAR, FLORIDA 32950

Copyright © 1972 by
Applied Science Publishers Ltd.,
Transferred to J.M. Beer
and Norman Chigier
Reprinted by Arrangement

Printed in the United States of America

Library of Congress Cataloging in Publication Data

Beér, J.M. (János Miklós), 1923-
  Combustion aerodynamics.

  Reprint. Originally published: London:
Applied Science Publishers, 1972. (Fuel and
energy science series)
  Bibliography: p.
  Includes index.
  1. Combustion. 2. Aerodynamics. I. Chigier,
N. A. II. Title. III. Series: Fuel and energy
science series (London, England)
QD516.B416 1983   541.3'61   82-13084
ISBN 0-89874-545-4

Contents

Preface

This book is intended as material for undergraduate and postgraduate courses on combustion, and also as a work of reference for practising engineers. Combustion is one of the oldest branches of science and, as in other areas of science and technology, rapidly rising interest in a wide range of applications has added a wealth of new knowledge, first during the period between the two world wars, and, in particular, in the last two decades. The large increase in the number of books, periodicals and conference proceedings has led to a high degree of specialisation within the broad spectrum of combustion technology, embracing branches of many sciences such as physics, chemistry, spectroscopy, thermodynamics, and also the engineering science of transport processes.

Our purpose is to discuss aerodynamic processes, which play important roles in turbulent diffusion flames, the type most generally used in industrial practice. A large proportion of the material in the book has been selected from the area of fully separated flows; turbulent jets and wakes with and without combustion. Whilst attempting to give a general picture, the limitations of a short monograph for presenting a comprehensive treatment had to be recognised. There is a preponderance of topics and discussion of results originating from our research work carried out at the Research Station of the International Flame Research Foundation at Ijmuiden, and later at the University of Sheffield. The convenience of following a line of thought familiar to us, rather than any other consideration, has led to this choice of material.

An attempt has been made to balance descriptive and analytical treatment in the monograph, and analysis, whenever possible, has been associated with a physical model of the process. Where a detailed mathematical analysis leading to a generalisation of results has been available it has been included, while in other cases a simpler, quantitative treatment based upon relationships between appropriate non-dimensional groups has been recommended as the most reliable tool for use by the designer. In this respect the treatment of the material reflects the authors' views of the 'state of the art'. The ideal is surely that the designer should be able to solve his problem by computation only, *i.e.* without need for resorting to experimentation. Prediction methods have made encouraging progress, and they will provide the guidelines for research in the years to come. But the designer will have to rely for some time ahead on semi-empirical relationships and on physical modelling. It is therefore equally important that this well-known approach to design should be further developed. Discussions have been included on both the analytical and critical modelling of combustion aerodynamics.

The authors have been given invaluable assistance by colleagues, associates, research students, technical assistants and secretaries. Many of the names of these helpers appear in references to papers given in the book. We should like to express our particular gratitude to Dr D. G. Lilley (Chapter 4), Dr A. S. Nuruzzaman (Chapter 6), Dr N. Syred (Chapters 3 and 8), Dr C. G. McCreath and Mr K. Bassindale (Chapters 3 and 8) and to members of our research team at the University of Sheffield, for their direct and valuable contributions to various chapters of the monograph. Our thanks also go to our secretaries Mrs J. Czerny, Mrs A. Thornhill and Miss S. Bonsall for typing and organising the material for the book, and to Mr P. G. Gillard and Mr R. Thompson for the high quality drawings and photographs.

We acknowledge with thanks the financial support of the Science Research Council, the Admiralty and Shell Research Ltd, for research work to which we have referred in the book.

J. M. Beér

January 1972   N. A. Chigier

CHAPTER 1

Introduction

The phenomenon of flame is the result of complex interactions of physical and chemical processes whose study involves numerous disciplines such as thermodynamics, chemical kinetics, fluid dynamics, etc. Although combustion is in general terms an exothermic oxidation reaction, physical processes, in particular transfer of energy of mass and of momentum, play significant roles in a combustion system and are the most important rate determining steps in the overall reaction of most industrial flames.

For the sake of discussion it is convenient to distinguish between premixed and diffusion flames. In premixed flames the reactants are mixed completely before entering the flame, while in diffusion flames fuel and air are not fully mixed or are not mixed at all before combustion starts. As a result, mixing of the reactants and combustion are concurrent in the latter type of flame.

In both premixed and diffusion flames the flow can be laminar or turbulent. The combustion technologist is primarily interested in

(a) the rate of propagation of the burning zone into the unburned mixture,

(b) the volumetric combustion rate in the flame, and

(c) the energy transfer, mostly by radiation, but also by convection from the flame to the surroundings.

Information on the mechanism and the rate of flame propagation will give flame stability limits. Data on the rate of combustion in the flame are necessary for determining the physical dimensions of the flame. The rate of energy transfer is important because it is the objective of some combustion systems to transfer as much heat as possible to a heat sink while it is important in others (gas turbine combustors) to avoid excessive heat transfer from the flame to the bounding surfaces. In either case details of the heat transfer process need to be known for efficient design and operation.

In both the processes of propagation of the burning zone and the

1

burnout of fuel in the flame, transfer of heat, mass and momentum play important roles. The transfer will be molecular in laminar flames and will be by turbulent diffusion in turbulent flames. In all cases, however, the rate of molecular or turbulent transfer will be dependent on the flow pattern in the flame or the combustor. The connection between mixing, *i.e.* concentration distribution on the one hand and flow pattern on the other, gives combustion aerodynamics its significance in flame studies.

Burners are devices for the injection of fuel and oxidant and their mixing in combustion chambers. Gaseous fuels can be pumped directly to burners for mixing with the oxidant, but liquid and solid fuels require preparation. Liquid fuels are normally heated in order to reduce their viscosity for efficient atomisation in the burner. The atomised fuel is then injected in the form of a cloud of finely dispersed droplets. Solid fuel can be burned on grates, in fluid beds or as pulverised fuel. Our discussion of burner flames will, however, be relevant only to the latter form in which fuel ground in mills is pneumatically transported to burners and injected into the combustion chamber as a cloud of finely dispersed particles. The oxidant most generally used is air which is pumped through the burner into the combustion chamber by means of a fan, blower or compressor. Fuel and oxidant are rarely completely premixed in the burner; they are either injected separately or more often the fuel is mixed with a small proportion of the combustion air only (primary air) on its path through the burner. The splitting of the combustion air into primary and secondary streams is made for reasons of flame stability and safety, such as prevention of flash back and explosion in the fuel supply lines. It also enables higher preheat temperatures to be used in the secondary air which in turn may favourably affect the thermodynamic efficiency of the system.

With the exception of those cases where the air stream fills the total cross section of the combustion chamber (*e.g.* cement kilns), the streams of fuel and air issuing from the burner are both in the form of a jet, *i.e.* flows fully separated from walls. The jet momentum of the fuel flow and the air stream are utilised for directing the flame and for controlling mixing in the combustor. The flow and mixing patterns in the flame are dependent upon the pressure energy that is converted into kinetic energy at the burner exit and also upon the burner geometry (Chapter 2).

Round jets are formed by fluid issuing from pipes or nozzles of circular cross section. Because of their symmetry about the jet axis, round jets may be considered 'two dimensional' for purposes of analysis, in a cylindrical co-ordinate system. For very low flow rates and small burner diameters the jets may be laminar at exit and subsequently break down into turbulent flow, but in general the size of industrial burners and the exit velocities are sufficiently large for the flow to be turbulent at nozzle exit.

Annular jets frequently serve for the introduction of secondary air around the primary jet that carries the fuel or surrounding an oil gun situated on the burner axis. Non-streamlined, so-called 'bluff bodies' are often placed in the centre of round nozzles. The flow is then issuing from an annular nozzle and a region of reverse flow forms in the wake of the bluff body which is instrumental to flame stabilisation.

Double concentric jets are compound jets consisting of a central and a coaxial annular jet. At a distance of several nozzle diameters downstream from the burner, the two jets combine and the behaviour of the compound jet in this region can be predicted with good approximation from the combined mass flow rates and jet momenta. Near the burner, however, size and geometry of the interface separating the central and annular nozzles has an important influence on the mixing. Because of its effect on flame stability, this region is of special interest to combustion technologists.

Jets penetrating into a main stream at some angle—*transverse jets*—have a frequent application as secondary air or dilution air injection into the flame (*e.g.* gas turbine combustors). For design purposes, knowledge of the path of the transverse jet, its penetration into and mixing with the main stream is necessary. While a complete analytical treatment of transverse jets is not possible, a semi-empirical approach to the prediction of significant parameters of the flow is shown to yield results with good approximation.

When jets are *confined* there is a limited supply of fluid available for entrainment. As a result, an adverse pressure gradient is set up along the jet associated with a recirculating flow outside the jet stream. Two methods of predicting characteristics of enclosed jets are presented. One of these is based on the simplifying assumption that the development of an enclosed jet, such as that of a free jet, is determined by its momentum and that the entrainment and spread of the jet is unaffected by the enclosure. The more comprehensive method is based on the detailed treatment of the equations of motion. It is shown that, while this second method gives more generally applicable results, the simple method can be used for most practical cases to predict the rate of recirculation and the position of the core of the re-circulation eddy from input parameters of the enclosed jet system.

As a result of combustion, density gradients arise in combustion chambers. The rate of entrainment of gases from the surroundings into a jet or flame is dependent upon the density differences between the jet and its surroundings. When jet densities are higher than those of the surrounding gases, entrainment is reduced and, conversely, a low density or high temperature jet penetrating into cold environment has a higher entrainment rate than a jet in a constant density system. These effects are taken into account when the combustion length and the concentration distribution in diffusion flames are being discussed.

Some basic concepts of flame theory are discussed in Chapter 3. In premixed fuel–oxidant mixtures the flame propagation can be described in terms of transfer of heat and chemically active species from the reaction zone upstream into the fresh mixture where, again, combustion reaction is initiated. The new layer then becomes the new source of heat and of chemically active species. In laminar flames the transport is by molecular diffusion, while under turbulent conditions the transfer of heat and mass is considerably augmented resulting in an increased rate of flame propagation. This can be further increased by convective flows such as occur in recirculation zones.

Recirculation zones are formed in flows when an adverse axial pressure gradient exceeds the kinetic energy of fluid particles and a stagnation point is formed. This can be brought about by (a) throttling entrainment to a turbulent jet by confinement in a chamber, (b) introducing a bluff body into the main stream, and (c) imparting strong swirl to jets.

Significant relationships exist between the size and strength of the recirculation vortex and the stability characteristics of the flame. On the other hand, the size and strength of the vortex will depend upon input conditions such as blockage ratio and forebody geometry. Some of the information is in a semi-empirical form where the relationship between relevant groups of parameters is determined experimentally.

An analysis of burning, swirling jet flows without recirculation is presented in Chapter 4. Prediction of time–mean average velocity, pressure, concentration and temperature in turbulent flames can be made provided the turbulent momentum flux tensor τ, the turbulent enthalpy flux vector $\mathbf{J}_h$ and the turbulent chemical species' flux vector $\mathbf{J}_j$ (one for each chemical species) are specified. Turbulent exchange coefficients μ, Γ_h and Γ_j (defined by analogy with Newton's, Fourier's and Fick's laws for laminar flows) are generally used and assumed to be isotropic, the latter two bearing fixed ratios (Prandtl and Schmidt numbers) to the former. A method is presented which allows the distributions of μ_{rz}, $\mu_{r\theta}$, $(\Gamma_h)_r$ and $(\Gamma_j)_r$ (the significant flux components in non-recirculating, swirling flames) to be determined from experimental time–mean distributions of v_z, v_θ, T and m_j. Calculations show that previous assumptions are not generally valid and that for swirling flows the turbulent stress distribution is non-isotropic. The exchange coefficients are shown to be functions of the degree of swirl and position in the flow field.

In swirling flows (Chapter 5) the fluid emerging from the orifice has a tangential or swirl velocity component in addition to the axial and radial components of velocity encountered in free, axial, non-swirling jets. Because of their wider spread and their faster mixing with the surrounding fluid, swirling jets are frequently applied to flames as an effective means of controlling mixing of reactants and of recirculating flow. Recirculating flow, which arises in the central region of strongly swirling jets, may be used to stabilise flames on burners. Practical methods of swirl generation are discussed together with the quantitative characterisation of swirl intensity. The non-dimensional ratio of two invariants, that of the angular and linear momenta of the jet, can be used as a measure of the degree of swirl. The stability of swirling flow in divergent nozzles—conditions of separation of fluid flow from the divergent nozzle walls—is considered for varying swirl intensities and for different nozzle geometries.

Because of the effective control of the proportion of nozzle fluid mass recirculated in swirling jets, the residence time distribution of fluid particles in a combustion chamber with such a jet can be varied over a wide range by the burner geometry and the swirl intensity of the flow. If the dependence of the overall rate of a combustion reaction upon the concentration of the reactants and temperature, together with the residence time distribution of the reactants in a combustion chamber, is known, good approximate calculations of combustion performance can be made. The ability to vary residence time distributions is shown to be valuable for optimising combustion performance.

When rotating flow is coupled with a strong positive density gradient in the radial direction, this can result in a damping of the turbulent exchange of mass and momentum in the radial direction and a corresponding increase in the combustion length of a turbulent jet flame burning along the axis of a rotating environment. A dimensionless criterion is given for the laminarisation of the boundary layer.

Heterogeneous combustion, the burning of droplet sprays, is discussed in Chapter 6. The kinetics of combustion of single droplets, of monosize drop arrays and of polydispersed sprays are considered. For purposes of qualitative statements on the mechanism of droplet combustion, single droplets may be regarded as a microcosm of a spray. Theoretical studies have shown that the combustion rate of a droplet is determined by the rate of vaporisation, which in turn is controlled by the rate of heat transfer from the flame to the drop. Experimental studies have enabled the combustion rate to be given as a function of input parameters such as the droplet diameter, partial pressure of oxygen and temperature of the ambient gas. When fuel droplets burn in a turbulent jet diffusion flame, the aerodynamic parameters such as the momentum flux of the jet, entrainment into the jet and nozzle geometry become significant. Twin fluid or blast atomised oil flames and pressure jet oil flames are shown as aerodynamically contrasting types. In blast atomised oil flames the combined momentum flux of the oil and atomising agent flow is high and is therefore the factor determining the development of the flame. By contrast, in pressure jet oil flames the energy for mixing of fuel, oxidant and hot combustion products is mainly imparted to the airstream, and the momentum flux of the oil spray is relatively low. In these latter flames the oil spray characteristics such as drop size, spray angle and the spatial overlap of fuel spray and air flow pattern have significant effects upon the stability and combustion characteristics of the flame.

In Chapter 7, the merits and laws of physical modelling of combustion systems are discussed. While there is a good prospect that analytical prediction procedures will in the near future be developed to the stage where they can be used for purposes of design, physical modelling is at present a powerful tool of the engineer. It will most likely still be needed in the future for predicting the highly complex geometries usual in industrial combustors and furnaces.

Modelling laws are, at best, based on the same system of differential equations which constitute the starting point for analytical prediction procedures. For purposes of modelling, however, these equations need not be solved; it is sufficient that they be reduced to form relationships between dimensionless ratios of forces, energies or masses. The judicious choice of the dimensionless groups required as the basis of modelling criteria for a particular system has to be made with engineering insight. The combination of physical modelling with analytical procedures will in many practical cases lead to the fastest and least expensive method of predicting combustor and furnace performance.

The final chapter is concerned with measurements in flames. The objective of measurement may be to explore flames, to test theoretical predictions of flame properties or to monitor certain flame parameters for purposes of control. Also, flames are sometimes used as steady-state systems, convenient for determining material properties at high temperatures.

Several of the methods discussed in Chapter 8 were developed in conjunction with research on industrial size flames by the International

Flame Research Foundation and are currently in use in many research centres in industry.

The ability to obtain detailed information on velocity, gas and solid concentration and temperature distributions in flames enabled scientists and engineers to better understand flame phenomena, to generalise results of their studies for purposes of physical modelling and to test mathematical prediction methods.

More recently there has been a growing interest in measurements leading to physical input data for mathematical modelling procedures. This requires further development of techniques capable of determining time resolved velocity, concentration and temperature distribution in flames.

CHAPTER 2

Separated Flows

NOMENCLATURE

a	distance from nozzle exit to effective origin.
b	width of plane jet; jet boundary.
C	concentration; recirculation vortex centre.
Ct	Craya–Curtet number $= 1/(m)^{1/2}$.
d	nozzle or orifice exit diameter.
d'_0	equivalent nozzle diameter.
D_1	inner diameter of annular orifice.
D_2	outer diameter of annular orifice.
G	momentum flux.
h	height of plane jet.
H	ambient head.
l	length of pipe; reference width.
L	duct radius or half width.
$\dot{m}$	mass flow rate.
$\dot{M}$	cross stream mass flow rate.
m	parameter (eqn. 2.42).
N	upstream stagnation point recirculation eddy.
$\bar{p}$	mean static pressure.
$\bar{p}_0$	mean static pressure in surrounding fluid.
P	downstream stagnation point recirculation eddy.
q	excess flow rate.
Q	total flow rate.
r	radial co-ordinate.
$r_{0.5}$	radius where $\bar{u}/\bar{u}_m = 0.5$.
r_1	radius at jet boundary.
S	arc distance along deflected jet axis; cross sectional area.
$\bar{u}$	mean velocity in x (axial) direction.
$u' = (\overline{u'^2})^{1/2}$	root mean square of fluctuating velocity component in axial direction.
U	jet velocity.

8

$\bar{v}$	mean velocity in r (transverse) direction.
v'	fluctuating component of velocity in transverse direction.
V	velocity in direction Y.
V_0	cross stream velocity.
x	axial co-ordinate, distance from nozzle exit; distance in direction transverse to cross stream.
x_p	distance from nozzle exit to end of potential core.
x_0	distance from nozzle exit to apparent origin of jet.
x_f	distance from nozzle exit to beginning of fully developed region.
X	distance in initial direction of jet.
y	transverse co-ordinate; distance in direction of cross stream.
Y	distance in direction normal to X.

Greek Symbols

$\alpha_{0.5}$	jet angle determined from line $\bar{u}/\bar{u}_m = 0.5$.
ε	eddy viscosity.
η	$= y/l$.
θ	temperature difference between measuring point and surroundings; Thring–Newby parameter (eqn. 2.22).
θ'	modified Thring–Newby parameter (eqn. 2.26).
θ_0	initial angle between jet and cross stream.
λ	ratio of stream (or annular) to initial jet velocity $= \bar{u}_s/\bar{u}_0$; ratio of cross stream to initial jet velocity $= V_0/U_0$; $= l/L$ (Craya–Curtet).
ν	molecular kinematic viscosity.
τ	shear stress.
ξ	$= \int_0^x \sigma/L \, dx.$
σ	angular momentum (eqn. 2.35).
ψ	stream function.
ψ_0	dimensionless stream function.

Subscripts

0	initial value at nozzle exit.
a	annular; air.
c	centre line; central jet.
e	entrained.
i	instantaneous.
j	jet.

m maximum value at a particular x.
r recirculation.
s stream (external or annular); on axis of deflected jet.
x axial direction; at axial station x; in direction X.

2.1 JETS

When fluid emerges from a nozzle it interacts with fluid from the surroundings to form a jet. Jet flows are classified as fully separated flows
because, after separation from solid surfaces, the solid surfaces no longer

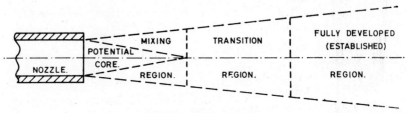

Fig. 2.1.   Regions of jet.

play a significant role in their development. Figure 2.1 represents regions
of turbulent free jets. Immediately downstream from the nozzle there is a
region, the potential core, within which the velocity and concentration of
nozzle fluid remain unchanged. Outside this region a free boundary layer

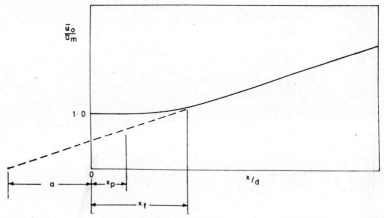

*Fig. 2.2.   Extrapolation of velocity on axis for determination of apparent origin of a
round jet.*

develops in which momentum and mass are transferred perpendicular to the direction of flow. The fully developed region of the jet is preceded by a transition region. The lengths of the potential core and transition regions are about 4 to 5 and 10 nozzle diameters respectively. These values also depend on initial conditions such as velocity distribution and turbulence level at the nozzle exit.

The fully developed regions of turbulent jets are similar, and therefore axial and radial distribution of velocities and concentration can be described by relatively simple and general relationships.[1-6]

Figure 2.2 represents the reciprocal of the axial velocity as a function of distance from the nozzle. The velocity at any point on the jet axis is independent of the nozzle diameter if the distance from the nozzle to the point is measured in terms of nozzle diameter. The equation corresponding to the straight line in Fig. 2.2 can be given as

$$\frac{\bar{u}_0}{\bar{u}_m} = 0 \cdot 16 \frac{x}{d_0} - 1 \cdot 5 \tag{2.1}$$

Because the mechanisms of transfer of momentum and mass are the same, a similar relation can be given for the axial distribution of concentration

$$\frac{C_0}{C_m} = 0 \cdot 22 \frac{x}{d_0} - 1 \cdot 5 \tag{2.2}$$

The horizontal part of the curve in Fig. 2.2 represents the distribution within the potential core and the difference between the slopes of the lines of velocity and concentration [see eqns. (2.1) and (2.2)] is due to the difference in the value of the coefficients for the turbulent transfer of momentum and mass respectively.

Because of the similarity of velocity profiles, the ratio of the axial velocity at any point to that on the axis at the same axial distance from the nozzle u/u_m is the same when given as a function of r/x, where r is the radial co-ordinate and x is the distance from the nozzle exit. This is true also for concentration distributions. The reduced or normalised velocity and concentration profiles are shown in Fig. 2.3 and the corresponding equations, assuming a Gaussian distribution, are

$$\frac{\bar{u}}{\bar{u}_m} = \exp\left[-K_u \left(\frac{r}{x}\right)^2\right] \tag{2.3}$$

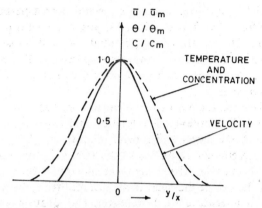

Fig. 2.3. *Dimensionless profiles of velocity, temperature difference and concentration for free jet.*

where K_u has values between 82 and 92 and, for concentration profiles,

$$\frac{C}{C_m} = \exp\left[-K_c\left(\frac{r}{x}\right)^2\right] \qquad (2.4)$$

where K_c has values between 54 and 57.

2.2 REPRESENTATION OF FLOW FIELDS

Flow fields may be represented in a number of ways. Velocity distributions can be given in the form of velocity profiles at a number of axial stations, lines of constant velocity (isovels), lines of constant relative velocity ($\bar{u}/\bar{u}_m$ = const) or streamlines may be drawn from integration of the velocity profiles. The velocity profiles in Fig. 2.4 show the initial conditions and the progressive changes which occur as the jet moves through the various regions. Profiles in this form are not convenient for describing flow fields in the downstream region where velocities are low. Lines of constant relative velocity referred to the maximum velocity at each particular cross section ($\bar{u}/\bar{u}_m$ = const) may be considered as rays emanating from the apparent origin, as shown in Fig. 2.5. These lines are used for the determination of jet angle, which is defined by the angle between the axis and one of these rays. $\alpha_{0.5}$ is the jet angle determined from the line $\bar{u}/\bar{u}_m = 0.5$ and $\alpha_{0.1}$ is determined from the line $\bar{u}/\bar{u}_m = 0.1$. The contours of a jet are represented by lines of constant velocity (isovels), as shown in

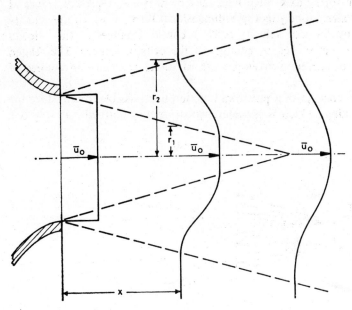

Fig. 2.4. Velocity profiles near jet exit.

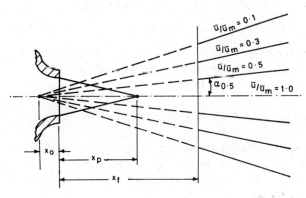

Fig. 2.5. Lines of constant relative velocity.

Fig. 2.6. These lines consist of a series of envelopes, all of which ultimately converge on to the axis. Such lines may be drawn to represent values of velocity, or alternatively in a non-dimensional form, in which the velocity is divided by the exit velocity ($\bar{u}/\bar{u}_0 =$ const). Changes in the velocity gradient are represented by changes in the density of lines. The related constant concentration envelopes are of direct relevance in studies of flames.

Using the concept of a potential function it is possible to introduce the stream function, which is a scalar, whose spatial derivative in any one

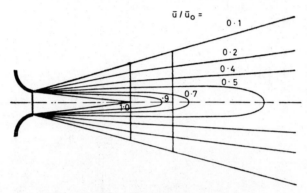

Fig. 2.6.   Equal velocity lines in submerged jet.

direction equals the component of the velocity vector in the same direction. The equation of a streamline can be given, in vector notation, as

$$\mathbf{U} \cdot d\mathbf{s} = 0 \tag{2.5}$$

At every point, the tangent to the streamline is the velocity vector. The stream function can be determined by integration of the axial component of velocity to various radial distances, as shown by the equation

$$\psi = \int_0^r \rho u r \, dr \tag{2.6}$$

From the radial distribution of stream function at a number of axial stations, streamlines can be drawn through points representing constant ψ. These values are usually non-dimensionalised by plotting lines of ψ/ψ_0 where ψ_0 is the integral

$$\psi_0 = \int_0^{d_0/2} \rho u r \, dr \tag{2.7}$$

as determined from the exit velocity profile. For jets with no change in density, these lines (Fig. 2.7) represent a series of annuli or stream tubes through which the mass flow remains constant, *i.e.* there is no net transfer of mass across the streamlines. These streamline patterns may be considered to represent what would occur physically if a stream of dye

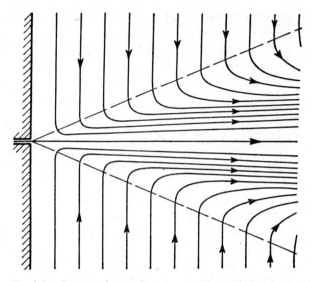

Fig. 2.7.   Pattern of streamlines in a circular, turbulent free jet.

were introduced between two streamlines. Because of the high rate of turbulent exchange, this dye would diffuse rapidly in the transverse direction, but the maximum concentration of dye would follow the streamlines as the flow continued downstream.

2.3   ENTRAINMENT

As a consequence of momentum exchange between the jet and surroundings, fluid is entrained from the surroundings across the boundaries of the jet. This entrainment is due to friction which depends upon the exchange coefficient and the velocity gradient. The turbulent exchange coefficient is several orders of magnitude higher than the molecular exchange coefficient. The magnitude of entrainment can be illustrated by the fact that a turbulent jet entrains fluid equivalent to the nozzle fluid mass flow rate for about every three nozzle diameters' distance along the jet axis. For a constant density system, entrained mass is given by[7]

$$\frac{\dot{m}_e}{\dot{m}_0} = 0.32 \frac{x}{d_0} - 1 \tag{2.8}$$

The more general relationship for non-constant density systems is

$$\frac{\dot{m}_e}{\dot{m}_0} = 0.32 \left(\frac{\rho_a}{\rho_0}\right)^{1/2} \frac{x}{d_0} - 1 \tag{2.9}$$

The momentum flux of a free jet is conserved and an expression can also be given for this.

If no external forces act on a free jet, then total momentum is conserved in every direction. When external forces are acting in any one direction, momentum is not conserved in that direction, but a force balance based on Newton's second law will show that all forces in that direction are balanced by the rate of change of momentum. The set of Reynolds' equations considers the force balance in each of the three directions in a particular co-ordinate system. If only part of the flow field is considered, turbulence stresses are present and momentum is not conserved. Also if there is any pressure variation in a flow field, momentum will not be conserved. For a free jet with no external forces acting, the total momentum flow per unit time in the axial direction is given by

$$\dot{G}_x = 2\pi \int_0^\infty \rho \bar{u}^2 r \, dr \tag{2.10}$$

At the nozzle exit

$$G_x = G_0 = \text{const} = \frac{\pi d_0^2}{4} \rho_0 \bar{u}_0^2 \tag{2.10a}$$

2.4   NON-CONSTANT DENSITY SYSTEMS

The general relationships which have been developed for single, free, round isodensity jets can be extended to more complex systems by using the concept of an equivalent nozzle diameter. In its simplest form, this concept is based on the conservation of momentum flux.

In a non-constant density system the jet entrains fluid of ambient density ρ_s and because of the high rate of entrainment the density of fluid within the boundaries of the jet approaches the density of the surroundings within a short distance from the nozzle exit. We can therefore write the momentum flux in the form

$$G_0 = \frac{d_e^2 \pi}{4} \rho_s \bar{u}_0^2 \qquad (2.10b)$$

This is equivalent to considering the fluid to emerge from a nozzle with diameter d_e with the same momentum and the same initial velocity, but with the density of the entrained fluid rather than the nozzle fluid. From the equations it follows that

$$d_e = d_0 \left(\frac{\rho_0}{\rho_s}\right)^{1/2} \qquad (2.11)$$

When the mass flow rate and the momentum flux (thrust) are known, the equivalent nozzle diameter can be calculated from

$$d_e = \frac{2\dot{m}_0}{(G_0 \pi \rho_s)^{1/2}} \qquad (2.11a)$$

2.5 JETS AND WAKES IN EXTERNAL STREAMS

JETS IN A CO-FLOWING PARALLEL STREAM
Now consider a system in which a jet is placed in a parallel stream of fluid in which the direction of the main stream and of the jet is the same. It is assumed that the thickness of the nozzle wall at exit is very small. Since turbulent mixing is dependent on velocity gradients, we can expect that, as the flow velocity of the external stream is increased, so velocity gradients and mixing between jet and external stream will be reduced until minimum mixing is reached when the velocities of jet and external stream are the same. When the velocity of the external stream is increased beyond that of the jet velocity, gradients are again set up and mixing is increased. Rates of jet spread and rates of decay of velocity and concentration will be reduced as the velocity gradients are diminished. Also the size of the potential core will increase until, for the case when jet velocity and external stream velocity are the same, the potential core extends throughout the flow field. This case has been studied theoretically by Squire and Trouncer[14] and experimentally for the axisymmetric case by Forstall and Shapiro[29] and more recently by Alpinieri.[30]

Squire and Trouncer[14] integrated the momentum equations using a mixing length and obtained numerical solutions for several values of the ratio λ, stream velocity to jet exit velocity. The axial velocity distribution in the mixing region near the jet exit is given by[14]

$$\bar{u} = \frac{\bar{u}_0 - \bar{u}_s}{2} \left[1 - \cos \pi \frac{r_2 - r}{r_2 - r_1} \right] \qquad (2.12)$$

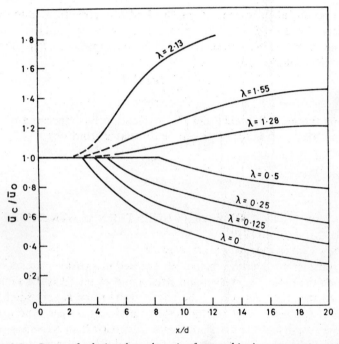

Fig. 2.8.   Decay of velocity along the axis of a round jet in a co-current stream.

where r_2 and r_1 are respectively the external and internal radii of the mixing region (see Fig. 2.4), $\bar{u}_s$ is the external stream velocity and $\bar{u}_0$ the initial jet velocity. Figure 2.8 shows the decay of velocity along the axis of a round jet in a co-current stream after the prediction of Squire and Trouncer for values of λ less than 1. Experimental results of Alpinieri[30] for λ more than 1 are also shown on the graph. It can be seen from Fig. 2.8 that the shortest potential core and fastest change of velocity are obtained for $\lambda = 0$ and $\lambda = 2.13$ and that, as λ increases from 0 to 1 or decreases from 2.13 to 1, the length of the potential core increases and the velocity decay decreases. Figure 2.9 shows the velocity lines, $u/u_m = 0.5$, for a round jet in a co-current stream as predicted by Squire and Trouncer for λ less than 1; it also shows the progressive decrease in jet angle and rate of spread as a function of λ.

Forstall and Shapiro[29] followed by Landis and Shapiro[32] verified experimentally that the predictions of Squire and Trouncer are, for most practical purposes, adequate for predicting the half boundaries of mixing.

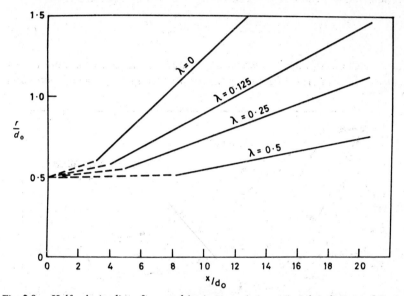

Fig. 2.9. Half velocity lines for round jet in a co-current stream (predictions of Squire and Trouncer[14]).

Their results showed clearly that property values in the mixing region are a function of the velocity ratio for either of the absolute velocities of the two streams. They have presented their results in terms of the length of the potential core which is given empirically, for $\lambda < 1$, as

$$\frac{x_p}{d_0} = 4 + 12\lambda \tag{2.13}$$

In the fully developed region, the velocity decay on the axis is given by

$$\frac{u_c - u_s}{u_j - u_s} = \frac{x_p}{x} \tag{2.14}$$

The rate of spread of the jet is given by

$$\frac{y_{0.5}}{r_0} = \left(\frac{x/d_0}{x_p/d_0}\right)^{(1-\lambda)} \tag{2.15}$$

where $y_{0.5}$ is the radial position at which the velocity is equal to

$$0.5(u_{min} + u_{max})$$

The velocity profile is given in terms of a cosine distribution as

$$\frac{u - u_s}{u_c - u_s} = \frac{1}{2}\left\{1 + \cos\frac{\pi r}{2y_{0.5}}\right\} \tag{2.16}$$

The studies of Alpinieri[30] were concentrated on the case of λ more than 1 because of the interest in applying the results to studies of rocket wakes and

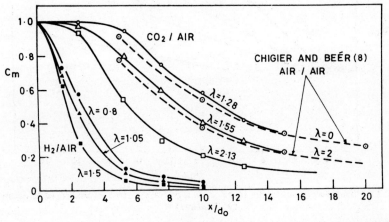

Fig. 2.10. *Decay of centre line concentration for round jet in a co-current stream (after Alpinieri[30]).*

ramjet combustors, where the central jet velocity is often smaller than that of the external stream. Alpinieri used carbon dioxide and hydrogen central jets exhausting into a moving concentric stream of air with flow velocities in the low to high subsonic range. The change of velocity along the axis is shown in Fig. 2.8 for a CO_2 jet in air and for $\lambda = 1.28$, 1.55 and 2.13. The decay of the centre line concentration for a round jet in a co-current stream is shown in Fig. 2.10. The decay of concentration along the centre line can be seen to depend both upon the stream to jet velocity ratio and the ratio of the densities. For the hydrogen jet in an air stream, the density ratio ρ_j/ρ_s is 2/29 and in this case the decay of concentration on the axis is so fast due to density differences that it is hardly sensitive to velocity ratios as shown in Fig. 2.10. For the CO_2 jet in air, for which ρ_j/ρ_s is equal to 44/29, concentration decay is much more sensitive to variation in the velocity ratio.

For cases in which there are both density and velocity differences between main stream and jet, the jet characteristics become a function of the ratio

$$\frac{\rho_s u_s}{\rho_{jo} u_{jo}}$$

This is borne out by the experiments of Alpinieri and also by those of Corrsin and Uberoi[33] who investigated the effect of large density differences by using a heated jet which exhausted into a quiescent region of different density. They found that a decrease of jet density with respect to that of the surrounding medium caused an increase in the rate of decay of the jet.

JETS IN A CROSS FLOWING STREAM

Jets issuing at some angle into a main stream undergo distortion so that they are not axisymmetric. The surrounding flow is decelerated on the upstream side of the jet creating a region of stagnation pressure while a low pressure region develops in the wake of the jet. Downstream of the jet in the main stream a pair of counter-rotating vortices form, similar to those in the wake of a cylindrical bluff body. The pressure difference between the upstream stagnation region and the 'wake' across the jet causes the jet to deflect and the jet cross section is distorted from a circular to a kidney shape.

The mixing between the jet and the main stream is accelerated by the additional lateral shearing stresses. This has the effect of shortening the potential core and the decay of velocity and concentration with distance from the nozzle is greater than it is for submerged jets.

Patrick[39] has carried out experiments in a low speed wind tunnel with stream velocities up to 8 m/sec and with jets penetrating at right angles into the main stream. Velocity and concentration distributions were determined from measurements made on the plane of symmetry of the system. By raising the air temperature by about 50°C in the nozzle the Schlieren technique could be used for flow visualisation. This optical method relies for its operation on the variation of the refractive index due to density gradients in the flow field under study.

Figure 2.11 represents the jet paths determined by Patrick[39] from

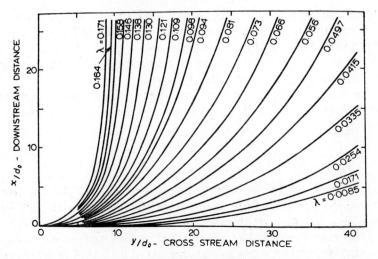

Fig. 2.11.   *Paths of jets in cross flowing stream;* $d_0 = 0.255$ in *(after Patrick[39]).*

Schlieren photographs. The parameter λ in the graph is the stream to jet velocity ratio. For non-constant density systems $\lambda^2 = (\rho_s/\rho_j)(V_0{}^2/U_0{}^2)$. From measurements of velocity and concentration the position of the jet axis was determined as[39]

$$\frac{y}{d_0} = 1{\cdot}0\lambda^{-0{\cdot}85}\left(\frac{x}{d_0}\right)^n \tag{2.17}$$

over the range of $0 \le \lambda \le 0{\cdot}152$.

The value of n in the above equation was found to be $0{\cdot}34$ and $0{\cdot}38$ for concentration and velocity data respectively.[39]

Both the decays of velocity and of concentration along the jet axis are given by Patrick[39] in terms of the dimensionless arc distance.

The velocity decay is

$$\frac{U_0}{U_m} = K_1 + K_2\left(\frac{\bar{S}}{d_0}\right) \tag{2.18}$$

where K_1 is constant and K_2 is a constant for a given value of λ. The axial decay of jet fluid concentration is given by the empirical relationship

$$\frac{C_0}{C_m} = \left[\left(\frac{\bar{S}}{d_0}\right)\exp\left(7{\cdot}8\lambda - 1{\cdot}85\right)\right]^{1{\cdot}18} \tag{2.19}$$

This relationship applies over the range $0 \leq \lambda \leq 0.152$ and for values of $\bar{S}/d_0$ and λ which give a value of C lower than its initial value of unity. The values of $\bar{S}/d_0$ for which $C \geq 1$ correspond to points within a 'concentration potential core'. The accelerated mixing between jet and surrounding fluid due to the cross flow is illustrated by Fig. 2.12 in which

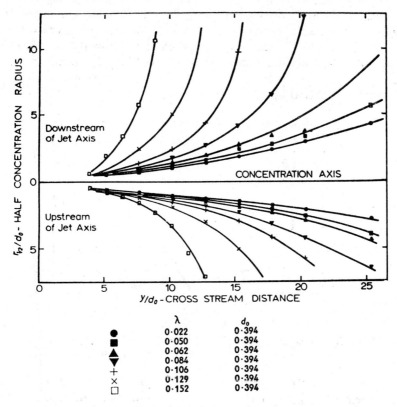

Fig. 2.12. *Spread of transverse jets.*

the spread of transverse jets is given as a function of the cross stream distance y/d_0. The boundary of the jet is defined by the half concentration radius of the jet, $r_{0.5}$, representing the distances in the direction of the main stream between the jet axis and points where the concentration has fallen to half the maximum value for the profile. For each cross stream distance y/d_0, two values of $r_{0.5}$ are given, one upstream and the other downstream of the jet.

2.6   ANNULAR AND COAXIAL JETS

In their fully developed regions (8–10 nozzle diameters downstream) the nozzle, annular and coaxial jets show flow patterns similar to those of round jets. There is, however, a displacement of the origin, *i.e.* the axial distance x must be replaced by $(x + a)$ in the equations giving velocity, concentration and entrainment along the jet. Close to the nozzle the annular jet sets up a region of underpressure with an associated closed ring vortex in the central region of the jet. Such vortices are also found in the wake of the annular interfaces between coaxial jets. The region close to the nozzle is therefore strongly affected by nozzle geometry of both annular and coaxial jets.

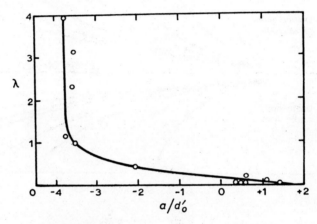

Fig. 2.13.   *The displacement of the effective origin in the hydrodynamic equations for the combined jet, as a function of the velocity ratio λ.*

Chigier and Beér[8] have determined experimentally values of the displacement of the origin of the jet for a wide range of values of the annular to central jet velocity ratio λ of coaxial jets (see Fig. 2.13). The value of the equivalent nozzle diameter d'_0 for a double coaxial jet system was calculated as

$$d'_0 = \frac{2(\dot{m}_c + \dot{m}_a)}{\left[\pi\rho(\dot{G}_c + \dot{G}_a)\right]^{1/2}} \tag{2.11b}$$

where $\dot{m}_c$ and $\dot{m}_a$ are the mass flow rates of the central jet and the annular jet respectively, and $\dot{G}_c$ and $\dot{G}_a$ are the respective momentum fluxes.

For the central jet alone, the characteristic velocity is the mean exit velocity u_{c0} through the central nozzle diameter d. The length of the potential core is dependent upon the shape of the nozzle and the intensity of turbulence generated before the exit from the nozzle. For the conditions of the same series of experiments,[8] Fig. 2.14 shows that for $\lambda = 0.08$ the velocity on the centre line remained constant for a distance of four central nozzle diameters. Thereafter, the axial velocity distribution tends towards the hyperbolic decay of

$$\frac{u_m}{u_{c0}} = 6.4 \frac{d'_0}{x + a} \qquad (2.1a)$$

As the annular exit velocity is steadily increased, Fig. 2.14 shows that the length of the central potential core is decreased and the decay of velocity is faster, until for the case of $\lambda = 2.35$ the central jet is completely absorbed by the annular jet so that at a distance of $3d$ reverse velocity is measured on the centre line.

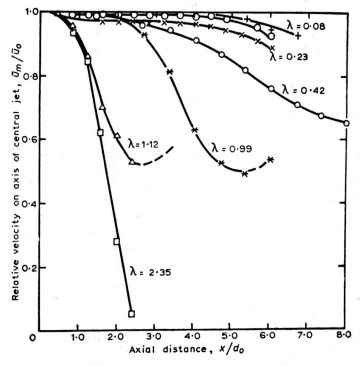

Fig. 2.14.   *Influence of the annular jet on the potential core and decay of the central jet.*

We may consider the annular jet in a similar manner by taking the annular width $\frac{1}{2}(D_2 - D_1)$ and the annular mean exit velocity u_{ao} as the characteristic length and velocity respectively. The potential core of the annular jet is not as easily distinguishable as that of the central jet but Fig. 2.15 shows that as the central jet velocity is increased, the length of the

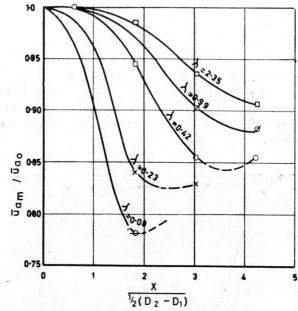

Fig. 2.15. Influence of central jet on potential core and decay of annular jet.

annular potential core is reduced and the decay of the maximum annular velocity is increased until, for the case of $\lambda = 0.08$, the annular jet is absorbed within the confines of the central jet after a distance of $(D_2 - D_1) = 1$.

2.7   CONFINED JETS

For an enclosed jet, such as a jet issuing into a duct, two extreme cases can be considered:

(a) there is ample supply of secondary flow surrounding the jet so that entrainment by the jet is uninhibited until it expands to reach the wall of the duct, and

(b) the surrounding secondary flow is less than that which the jet can entrain.

In this second case a recirculating flow is set up, *i.e.* some of the mean streamlines take the form of a closed loop as shown in Fig. 2.16. The details of this recirculating flow are of great interest to combustion engineers as the strength and size of the 'recirculation eddy' affect both

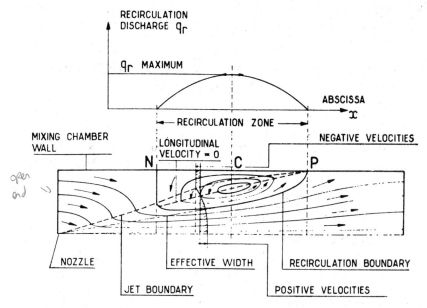

Fig. 2.16. Map of the recirculation for a ducted axisymmetric jet (after Barchilon and Curtet[42]).

the stability and the combustion length of turbulent diffusion flames. In the schematic representation of recirculating flow in Fig. 2.16, the secondary fluid is entrained upstream of point N and the recirculation eddy extends from the downstream boundary at point P to its upstream limit N. The rate of the reverse flow, defined as the integral of the negative velocities across the cross section, varies between these two points as shown at the top of Fig. 2.16. Its value q_r reaches a maximum at cross section C between two zero values at points N and P.

Thring and Newby[5] proposed a simple theoretical treatment of the problem based on the assumption that the rate of entrainment of the jet is unaffected by the enclosure and that the development of the jet is determined by its momentum flux. Craya and Curtet[41] and Barchilon and Curtet[42] have extended the theoretical treatment by making it more generally valid, but the simple generalisations made by Thring and Newby are still applicable to a wide range of practical cases; they give good agreement with experiment when the nozzle diameter is smaller than 1/10 of the duct diameter.

Field *et al.*[43] have used the entrainment equation by Ricou and Spalding,[7]

$$\frac{\dot{m}_e}{\dot{m}_0} = 0{\cdot}32 \left(\frac{\rho_a}{\rho_0}\right)^{1/2} \frac{x}{d_0} - 1 \tag{2.9}$$

for calculating the distance from the jet origin to the point P where the jet extends to the wall: assuming a jet angle α of 9·7°,

$$X_p = 5{\cdot}85L \tag{2.20}$$

where $2L$ is the duct diameter.

By treating the primary and secondary flows as a single jet (*i.e.* $X_N = 0$), the distance from the nozzle to point C in Fig. 2.16 can be given as

$$X_1 = \frac{1}{2}\left[X_p + \frac{d'_0}{0{\cdot}32}\left(\frac{\rho_0}{\rho_a}\right)^{1/2}\right] \tag{2.21}$$

By introducing a parameter

$$\theta = \frac{d'_0}{2L}\left(\frac{\rho_0}{\rho_a}\right)^{1/2} \tag{2.22}$$

eqn. (2.21) can be written as

$$X_1 = \frac{1}{2}\left(X_p + \frac{2L\theta}{0{\cdot}32}\right) = L\left(2{\cdot}925 + \frac{\theta}{0{\cdot}32}\right) \tag{2.23}$$

For the case where $\dot{m}_e = \dot{m}_r$, the fluid entrained between the nozzle and point C is given by

$$\frac{\dot{m}_e}{\dot{m}_0} = \frac{0{\cdot}32}{\theta}\frac{X_1}{2L} - 1 = \frac{0{\cdot}16}{\theta}\left(2{\cdot}925 + \frac{\theta}{0{\cdot}32}\right) - 1 = \frac{0{\cdot}47}{\theta} - 0{\cdot}5 \tag{2.24}$$

For the case when all the secondary flow is entrained by the jet before recirculation starts (Fig. 2.16), $\dot{m}_e = \dot{m}_a + \dot{m}_r$ and the proportion of recirculated mass flow is given as

$$\frac{\dot{m}_r}{\dot{m}_0 + \dot{m}_a} \tag{2.25}$$

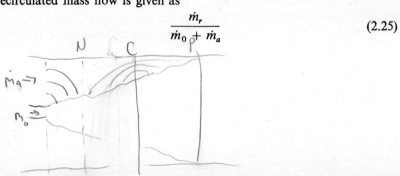

$$d_o' = \frac{2\dot{m}_o}{(G_o \pi \rho_a)^{1/2}}$$

$$\frac{1}{C_{\phi o}} = \frac{\dot{m}_a + \dot{m}_o}{\dot{m}_o}$$

and the modified Thring–Newby parameter as

$$\theta' = \frac{\dot{m}_a + \dot{m}_0}{\dot{m}_0} \frac{d_0}{2L} \left(\frac{\rho_0}{\rho_a}\right)^{1/2} = \frac{\dot{m}_a + \dot{m}_o}{(G_o \pi \rho_a)^{1/2} L} \qquad (2.26)$$

Substituting eqns. (2.25) and (2.26) into eqn. (2.24), we have

$$\frac{\dot{m}_r}{\dot{m}_0 + \dot{m}_a} = \frac{0\cdot47}{\theta'} - 0\cdot5 \qquad (2.25a)$$

Equations (2.24), (2.25) and (2.26) gave the position of points N and C in Fig. 2.16 as[43]

$$X_N = 6\cdot25\theta'L \qquad (2.27)$$

and

$$X_C = 3\cdot12(\theta' + 0\cdot94)L \qquad (2.28)$$

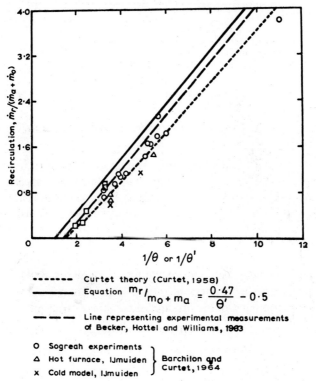

-------- Curtet theory (Curtet, 1958)

———— Equation $\dfrac{m_r}{m_0 + m_a} = \dfrac{0\cdot47}{\theta'} - 0\cdot5$

— — — Line representing experimental measurements of Becker, Hottel and Williams, 1963

O Sogreah experiments
△ Hot furnace, IJmuiden   ⎫ Barchilon and
x Cold model, IJmuiden   ⎬ Curtet, 1964
▫ Cold model, BCURA - Wingfield and Martin, 1966

Fig. 2.17.   Confined jet: maximum quantity of recirculation.

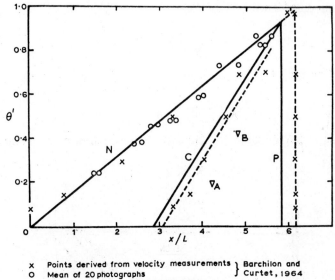

Fig. 2.18.   Position of recirculation eddy in confined jet.

Figures 2.17 and 2.18 illustrate the maximum quantity of recirculation and the position of the recirculation eddy in a confined jet respectively[43] as a function of the similarity criteria θ (Thring–Newby) and m (Craya–Curtet).

Craya and Curtet[41] developed an 'approximate confined jet theory' based on Reynolds' and the continuity equations. They made the following assumptions:

(a) The boundary layer approximation is valid for Reynolds' second equation giving a constant term $p + \rho \overline{v'^2}$ across the jet. Terms of acceleration and of longitudinal shear stress variation are negligible.

(b) Fluctuation terms in Reynolds' first equation are negligible.

(c) Similarity of the normalised excess velocity profile: $\Delta u = u - u_s$ (excluding wall boundary layer).

(d) No energy loss in the ambient flow, i.e. turbulence level in secondary stream is low.

On the jet axis $u = u_m$ and the excess velocity $\Delta u = \Delta u_m \doteq u_m - u_s$ (for notation see Fig. 2.19).

The similarity of the excess velocity profile is expressed as

$$\frac{\Delta u}{\Delta u_m} = f\left(\frac{y}{l}\right) \tag{2.29}$$

where l is a reference width dependent on axial distance x and its variation expresses the spreading of the jet. The width is defined by the expression

$$\pi l^2\, \Delta u_m = q = \int_0^b 2\pi y\, \Delta u\, \mathrm{d}y \tag{2.30}$$

where q is the excess flow rate and b is the jet boundary radius at which $\Delta u = 0$.

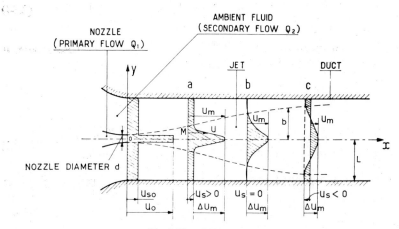

Fig. 2.19. *Main notation.*

The effective width l can be represented as the radius of the base of a cylinder of volume q and height u_m.

When the turbulence level in the ambient flow is low, the ambient flow can be considered as being potential flow and the ambient head can be given as

$$H_1 = \frac{\Delta p_1}{\rho g} + \frac{u_s^2}{2g} + \frac{v_s^2}{2g} \tag{2.31}$$

where Δp_1 is the pressure difference between the static pressure at the wall in a particular cross section and the pressure upstream in the ambient fluid, and u_s and v_s are the axial and transverse components of the mean ambient velocity at the same cross section.

The continuity equation for the total flow in the duct is then given by

$$\dot{Q} = \pi \, \Delta u_m l^2 + \pi u_s L^2 \tag{2.32}$$

where L is the duct radius.

The boundary layer thickness is neglected, and it is assumed that $d^2 \ll L^2$. The momentum equation reduces to

$$\frac{\partial (u_s^2 \, \Delta u_m l^2)/\partial x}{u_s \, \Delta u_m^2 l} + \frac{\partial (k \, \Delta u_m^2 l^2)/\partial x}{\Delta u_m^2 l} = 0 \tag{2.33}$$

where k is a form factor for the excess velocity profile

$$k = 2 \int_0^b f^2 \eta \, d\eta \tag{2.34}$$

where $\eta = y/l$.

A weighted mean of Reynolds' first equation across a cross section is then written by Curtet[42] in the form of a moment of momentum

$$\sigma(x) = - \int_0^b \frac{2\tau\eta}{\rho \, \Delta u_m^2} A \, d\eta$$

$$= -\alpha \frac{\partial (\Delta u_m^2 l^2)/\partial x}{\Delta u_m^2 l} + \frac{1}{4} \frac{\partial (\Delta u_m l^2)/\partial x}{\Delta u_m l} + \beta \frac{\partial (u_s l^2)/\partial x}{\Delta u_m l} \tag{2.35}$$

where A, α and β are form factors for the function f and are given as

$$\frac{1}{4A} = \int_0^b fF \, d\eta \tag{2.36}$$

$$\frac{\alpha}{A} = \int_0^b k(M - F) \, d\eta \tag{2.37}$$

$$\frac{\beta}{A} = \int_0^b f\eta^2 \, d\eta \tag{2.38}$$

where

$$F = 2 \int_0^\eta f\eta \, d\eta \tag{2.39}$$

and

$$kM = 2 \int_0^\eta f^2 \eta \, d\eta \tag{2.40}$$

For the case of $2L =$ constant (constant diameter duct), $U = \dot{Q}/\pi L^2$ and we can write $u_s/u = U_s$, $\Delta u_m/u = \Delta U_m$, $l/L = \lambda$ and $\sigma\, dx/L = d\xi$.

The following non-dimensional relationships are obtained

$$1 = U_s + \Delta U_m \lambda^2 \tag{2.41}$$

$$-\frac{3}{2}(\Delta U_m)^2\, \lambda^4 + \Delta U_m \lambda^2 + k(\Delta U_m)^2\, \lambda^2 = m \tag{2.42}$$

$$1 = -\alpha\,\frac{\partial[(\Delta U_m)^2\, \lambda^2]/\partial \xi}{(\Delta U_m)^2\, \lambda} + \frac{1}{4}\frac{\partial(\Delta U_m\, \lambda^2)/\partial \xi}{\Delta U_m\, \lambda} + \beta\,\frac{\partial(U_s\, \lambda^2)/\partial \xi}{\Delta U_m\, \lambda} \tag{2.43}$$

Equation (2.41) is the continuity equation, (2.42) is the momentum equation defining a constant parameter (the Craya–Curtet parameter) along the duct and (2.43) is the moment of momentum equation.

The Craya–Curtet Similitude Parameter

This parameter can be defined as above or it can be found directly by applying Euler's theorem to a surface including the duct and two of its

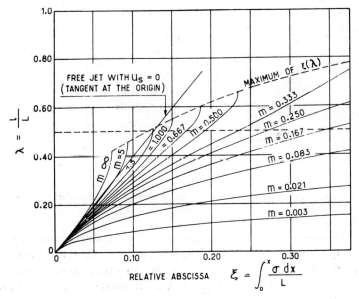

Fig. 2.20.   *Theoretical curves for non-dimensional effective width versus relative abscissa and parameter m.*

cross sections S; the following relationship can then be obtained (neglecting the fluctuation terms),

$$m + \frac{1}{2} = \frac{1}{U^2 S} \int \int_s \left(\frac{p}{\rho} + U^2 \right) dS$$

The parameter m can be expressed by the initial conditions, *i.e.* by the ratios of jet and ambient flow rates and of nozzle and duct radii.

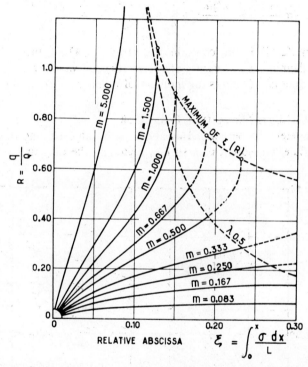

Fig. 2.21. *Theoretical curves for non-dimensional flow rate versus relative abscissa and parameter m.*

For the simple case when the nozzle to duct diameter ratio is small ($d \ll L$), the mixing conditions can be given as a function of m alone.

Curtet[44] presented the solutions of the non-dimensional system of equations in the form of $\lambda(\xi, m)$ and $R(\xi, m)$ curves. These are shown in Figs. 2.20 and 2.21.

Figures 2.22 and 2.23 show comparison of measured and calculated

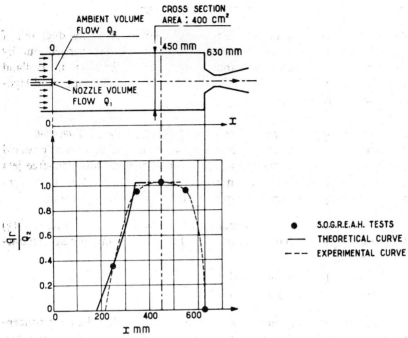

Fig. 2.22. *Experimental and theoretical curves showing recirculation flow rate variation in terms of abscissa.*

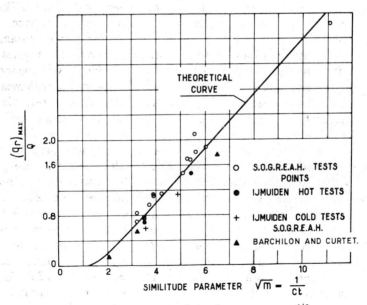

Fig. 2.23. *Maximum recirculation flow rate versus $m^{1/2}$.*

flow rates of recirculation as a function of distance along the duct and of the similarity criterion m. The experimental work was carried out by the International Flame Research Foundation at Ijmuiden in Holland and by SOGREAH (Société Grenoblois d'Aerodynamique et d'Hydraulique) in France.

COMPARISON OF THE THRING–NEWBY AND CRAYA–CURTET THEORIES

The Thring-Newby theory[5] is based upon the simple notions of physical behaviour of free and ducted jets. It assumes that all turbulent-free jets are dynamically similar. This in turn permits the use of $u_m/u_0 \times L/r_0$ and $C_m \times L/r_0$ as dependent variables. It is further assumed that

(1) the radial spread functions $C/C_m = f_1(\eta)$ and $u/u_m = f_2(\eta)$ (where $\eta = r/x$) are dependent only on the Reynolds number at a characteristic point,

    (2) momentum is conserved, and

    (3) nozzle fluid mass is conserved.

For enclosed jets, they stated as a further condition that the mass concentration of nozzle fluid after complete mixing is $C'_\infty = m_0/(m_0 + m_a)$, and therefore for similarity $C'_\infty (L/r_0)$ must be equal in model and in prototype.

From the point of view of combustion engineering design, the Thring–Newby criterion has advantages over the Craya–Curtet number. The value of θ can simply be determined from the values of input parameters (mass flow rates and radii of nozzle and duct respectively). Figures 2.17 and 2.18 show that calculations based on the Thring–Newby theory give good agreement with those of the Craya–Curtet theory and with measurement data. This is particularly so for cases when $(r_0/L) < 0.1$. Also the Thring–Newby theory predicts the position of the downstream stagnation point P which cannot be predicted by the Craya–Curtet theory.

The Craya–Curtet theory is more rigorous and more generally valid. More detailed predictions can be made for flow conditions using this theory than from the Thring–Newby theory. It is also a good example of theoretical analysis carried through from the Reynolds equations of motion to provide predicted data for the case of ducted jet that has shown to agree well with experiment. In contrast to more recent prediction procedures, the Craya–Curtet theory describes the flow field by means of significant parameters such as the similitude parameter, the magnitude and the position of the recirculation, the spread of the jet, etc. Also this method requires much less computer time for the solution than does a zone method or a finite difference method of integration of the equations.

The disadvantage of the Craya–Curtet theory is that it is mathematically complex.

Barchilon and Curtet[42] calculated values of the Craya–Curtet number and those of the Thring–Newby criterion for a series of their experiments. These values are tabulated below together with the input data of the experimental runs.

Test no.	Jetting flow Q_1 litres/sec	Ambient flow Q_2 litres/sec	Thring–Newby criterion θ	Craya–Curtet number $Ct = (1/\sqrt{m})$
1	1·463	16·89	0·941	0·976
2	1·469	12·14	0·695	0·714
3	1·418	7·96	0·496	0·506
4	1·450	4·37	0·301	0·305
5	1·466	1·48	0·151	0·152
6	1·456	0	0·075	0·075

As can be seen from the comparison of values of θ and of $m^{-1/2} = Ct$, the agreement is very good at low values of θ ($\theta < 0.3$) and the difference is still within 5% as the value of θ approaches unity.

2.8   JET FLAMES

Changes in flame shape can be observed when the flow rate of fuel gas issuing from a round nozzle is progressively increased. When fuel gas discharges at velocities below a critical value from the nozzle into stagnant air surroundings, the flow of gas is laminar and the mixing of gas and air occurs by molecular diffusion in a thin flame surface which is fixed in space. As the nozzle velocity is increased, the diffusion flames increase in length until a critical velocity is reached and the tip of the flame becomes unsteady and begins to flutter. With further increase in velocity, this unsteadiness develops into a noisy turbulent brush of flame starting at a definite point along the flame where breakdown of laminar flow occurs and a turbulent jet develops. The distance from the nozzle to the point where the turbulent brush begins is termed the breakpoint length. Characteristic flame length and breakpoint length curves as initially determined by Hottel and Hawthorne[4,46] are shown in Fig. 2.24.

As the nozzle velocity increases from zero, initially there is an almost proportional increase in flame length and, at any velocity in this region, the flame is sharp edged and constant in shape. When a sufficiently high velocity is reached, the tip of the flame changes in character and a slight brush forms. With further increase in velocity, the point at which the flame 'breaks' moves towards the nozzle and, with some fuels, the flame length is slightly reduced. When the breakpoint has approached quite

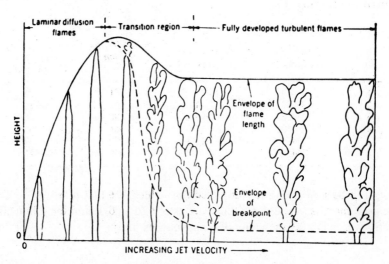

Fig. 2.24. *Change in flame length with increase in nozzle velocity (Hottel and Hawthorne[46]).*

close to the nozzle, the fully developed turbulent flame condition is reached. Further increase in velocity has practically no effect on flame length, but flame noise continues to increase and flame luminosity continues to decrease. The flame finally blows off the nozzle at a velocity depending on the type of gas and on nozzle size.

2.9   LAMINAR JET FLAMES

The first complete theory of laminar diffusion flames was developed by Burke and Schumann in their classical paper.[45] They solved the two-component diffusion equations for the position of the flame front determined as the loci of material sink in the equations. Later Hottel and Hawthorne[46] presented a simplified semi-empirical treatment of laminar jet flames and showed that the length of laminar flames can be calculated from

$$L = A \log_{10} Q\theta_f + B \qquad (2.44)$$

where L is flame length, Q is volumetric flow rate, A and B are constants independent of velocity and nozzle diameter and θ_f is a time factor that can be calculated from

$$\theta_f = \frac{1}{4 \ln [(1 + a_t)/(a_t - a_0)]} \qquad (2.45)$$

where a_0 is the primary air/fuel gas molal ratio in the nozzle fluid and a_t is the stoichiometric ratio. The values of constants A and B vary with type of gas and with a_0. For city gas with $a_0 = 0$, values of $A = 1.39$ and $B = 5.09$ were found for L and Q having dimensions feet and cubic feet per second at room temperature and pressure.

The breakpoint was found to be perfectly stable and there was no random breakdown from laminar to turbulent flow caused by disturbances in the air, except with very long diffusion flames from small nozzles in the region of the critical velocity.

The behaviour of a liquid jet was found to be very similar to that of flames in the transition region, suggesting that the phenomenon is related to fluid flow rather than to combustion. Transition from laminar to turbulent flow is usually indicated by the Reynolds number. The Reynolds number is inversely proportional to the kinematic viscosity and the value of the viscosity increases drastically in the combustion region (a factor of 15 to 20). Thus turbulent flow in the burner, characterised by a high Reynolds number, may give rise to laminar flow within the flame resulting in tall cylindrical flames where mixing takes place by molecular diffusion rather than by turbulent entrainment. Values of Reynolds number (cold) at which transition from laminar to turbulent flames commenced were given as:

		Reynolds Number $\times 10^{-3}$
1.	Hydrogen (zero primary air)	2
2.	City gas (zero primary air)	3–4
3.	CO (zero primary air)	5
4.	Hydrogen (with primary air)	5·5–8·5
5.	City gas (with primary air)	5·5–8·5
6.	Propane, acetylene (zero primary air)	9–10
7.	Methane	3

2.10   COMBUSTION IN TURBULENT JET DIFFUSION FLAMES

Measurements of visible flame length depend on the decision of the observer and conditions of observation. Long-time exposure photographs give flame lengths within 10% of those determined from the location of the point of 99% complete combustion, as determined by gas analysis.

The length of a turbulent diffusion-free flame is mainly dependent upon the type of fuel used and upon the diameter of the burner orifice. Guenther[47] has found the following semi-empirical formula:

$$\frac{L}{d} = 6(R + 1)\left(\frac{\rho_e}{\rho_F}\right)^{1/2} \tag{2.46}$$

where R = stoichiometric air/fuel weight ratio (17·25 for CH_4);

ρ_e = fuel gas density   0·0423 lb/ft^3 for CH_4
                                  0·677 kg/m^3 at 15°C;

ρ_F = mean flame density 0·0128 lb/ft^3 at 3000°R
                                  0·205 kg/m^3 at 1400°C.

The flame density is about the same for most gaseous fuels. Therefore, the ratio L/d depends only on R and ρ_e. For natural gas

$$\frac{L}{d} = 6(17·25 + 1)\left(\frac{0·0423}{0·0128}\right)^{1/2} = 199$$

The formula has been found to be accurate within 10%. A value of $L/d = 200$ is recommended for free natural gas flames. Values of L/d for other fuels are:

Fuel gas	L/d
CO	76
Coke oven gas	110
City gas	136
H_2	147
C_2H_2	188
C_3H_8	296
CH_4	200

Hawthorne et al.[4] made the simplifying assumptions of constant temperature concentration and velocity across a cross section within jet flames, constant angle of jet spread and chemical reaction occurring as soon as mixing takes place. By equating the buoyancy force to the change in momentum they showed, analytically, that the air entrained in a burning jet is initially less than in an isothermal jet. When combustion is completed, the temperature thereafter decreases with further mixing, resulting in a progressive change in the rate of entrainment.

In their experiments on fuel gas flames, Hawthorne et al.[4] showed that, as the exit velocity is increased above the critical value, the flame lengths and breakpoint lengths attain a constant value or one which varies only slightly with velocity. When the fuel gas is premixed with air, the length of the fully developed turbulent section is shorter. At certain nozzle velocities, either the normal on-burner flame or an off-burner flame can be produced. In the latter, the flame is not attached to the nozzle and combustion begins above the nozzle at a point which is relatively stable. On-burner city gas flames are luminous while off-burner flames are blue and non-luminous. Both on- and off-burner propane flames are luminous. If s is the distance from the nozzle to breakpoint, fully turbulent flame lengths $(L - s)/d$ are relatively independent of nozzle diameter and of nozzle velocity.

2.11 CONCENTRATION DISTRIBUTION IN JET FLAMES

The results of concentration measurements by Hottel and Hawthorne[46] in hydrogen flames are a good example of the fact that mixing in flames cannot be explained simply in terms of density differences between flame and surroundings. They found that decay constants are very much higher in flames than in isothermal jets, despite the fact that the densities in the flames were lower due to the lower molecular weight of the hydrogen and also the higher temperatures in the flame.

A number of non-burning studies of entrainment in which jet densities differ from that of the surroundings show that

$$\frac{C_m}{C_0} = K_c \left(\frac{\rho_j}{\rho_s}\right)^{1/2} \frac{d}{x + a} \qquad (2.47)$$

where ρ_j and ρ_s are initial densities in the jet and surrounding fluids and K_c is the decay constant for an isodensity jet. If we take the value of K_c to be 5·8, then for non-burning hydrogen jets in air we should have

$$K'_c = 5 \cdot 8 \left(\frac{2}{29}\right)^{1/2} = 1 \cdot 5 \qquad\qquad (2.48)$$

i.e. the decay rate of a hydrogen jet in air should be considerably greater than that in a constant density system. The measurements of Hottel and Hawthorne on hydrogen flames burning in air gave the following values of K_c:

Primary air/gas ratio	K_c
0	50
0·2	26
0·4	16
0·6	16

Thus the decay rates of hydrogen flames are much smaller than those for non-burning hydrogen jets in air. This can be partly explained by the position of the reaction zone in diffusion flames. The reaction zone forms an envelope around a cold core of gas emerging from the nozzle and the flame can therefore be considered as a cold, high density jet entraining hot, low density combustion products. This then results in little entrainment and a low rate of decay. This was borne out by the evidence of Hawthorne[4] who found that approximately eight times as much length was required for mixing in hydrogen flames as in an air jet.

Hawthorne[4] showed that concentration profiles in flames may be shown to be the same as those in isodensity jets by plotting C/C_m against $r/r_{0 \cdot 5C_m}$, where $r_{0 \cdot 5C_m}$ is the value of r at which $C = 0 \cdot 5C_m$. This manner of plotting profiles does not reveal much information concerning differences in profiles since all profiles are forced to pass through the two points at $C/C_m = 1$ and $0 \cdot 5$.

Examination of the rates of spread of jets and of jet flames is also revealing. Hawthorne measured $\alpha_{0 \cdot 5C_m} = 3 \cdot 1°$ for hydrogen diffusion flames with $(a/d) = 40$ to 50. Chigier and Chervinsky[19] measured $\alpha_{0 \cdot 5U_m} = 2 \cdot 5°$ for petroleum gas flames with $(a/d) = 35$. If these values are contrasted with measurements made in isodensity non-burning systems of $\alpha_{0 \cdot 5U_m} = 5°$ and $\alpha_{0 \cdot 5C_m} = 7 \cdot 5°$ with $(a/d) = 2$, it can again be clearly seen that turbulent diffusion flames have a significantly lower rate of spread than isodensity non-burning jets and that, as a result of combustion, effective origins are displaced from $(a/d) = 2$ to $(a/d) = 40$ upstream from the nozzle exit.

The changes brought about due to combustion can be observed in off-burner flames. Axial concentration measurements in an off-burner propane flame showed that the initial stage of mixing is strikingly different from that exhibited by a flame attached to the burner. The mixing rate between the burner and the flame has the higher value associated with an isothermal jet while, after ignition, the mixing rate changes to the lower value associated with a burning jet.

Kremer[48] has found that both the profiles of mass flux density and of momentum flux density maintain smooth profiles across the mixing zone. They also remain similar at all axial stations. Furthermore, these profiles remain unchanged whether or not the jet stream is burning. The experimental results of Kremer[48] are correlated by the equations

$$\text{Mass flux density:} \quad \frac{\rho u \chi}{(\rho u \chi)_m} = \exp\left(\frac{-\ln 2y}{y_{0.5}}\right)^2 \qquad (2.49)$$

$$\text{Momentum flux density:} \quad \frac{\rho u^2}{(\rho u^2)_m} = \exp\left(\frac{-\ln 2y}{y_{0.5}}\right)^2 \qquad (2.50)$$

where $y_{0.5}$ is that jet radius where the velocity is half of the maximum found on the axis at the same x and χ is the mass concentration of nozzle gas.

Temperature profiles have maxima away from the axis along the flame envelope and temperatures on the axis increase as a result of mixing with combustion products diffusing away from the flame. The temperature on the axis thus increases until it reaches a maximum when the flame envelope has converged onto the axis. Thereafter, temperatures decrease as the flame becomes cooled by mixing with the surrounding air.

Comparison of Angles and Effective Origin in Flames and Isothermal Jets

	$\alpha_{0.5U_m}$	$\alpha_{0.5C_m}$	a/d	Investigators
1. Isothermal	4·8		2·3	Chigier and Chervinsky[19]
	5·5	7·9		Abramovich[21]
		6·45	7	Sunavala et al.[6]
2. Flame (i) Free hydrogen		3·1	40 to 50	Hawthorne et al.[4]
(ii) Free petroleum gas	2·5		35	Chigier and Chervinsky[19]

2.12 CONFINED FLAME LENGTH

When a free turbulent diffusion flame is confined in an enclosure, flame length increases due to the presence of recirculation, *i.e.* the depletion of the oxygen concentration in the gas entrained by the jet. This is dependent on a reasonable amount of excess air being supplied. If we consider

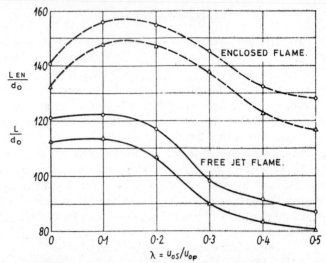

Fig. 2.25. *Combustion lengths of free and enclosed turbulent jet flames as a function of the secondary to primary air velocity ratio in the double concentric jet burner,* o *by visual observation,* Δ *99% combustion efficiency (Eickhoff and Lenze[49]).*

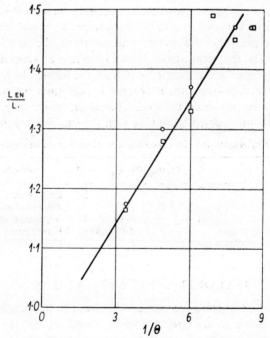

Fig. 2.26. *The ratio of enclosed to free jet combustion lengths as a function of the Thring–Newby recirculation parameter (after Eickhoff and Lenze[49]).*

the end of the flame to be where combustion is 99 % complete, the supply of the stoichiometric quantity of air only makes the flame infinitely long.

Experiments carried out by Guenther[47] for enclosed natural gas flames with 110 % of stoichiometric air showed $(L/d) = 250$.

Confinement not only results in increased length of the flame but also in widening of the flame.

The high values of CO_2 between the flame boundary and the wall of the combustion chamber are due to the recirculation gases which have a low concentration of oxygen. This results in a wider and longer flame, compared to a free flame, and a more uniform temperature profile.

Eickhoff and Lenze[49] have compared combustion lengths of free and enclosed turbulent jet flames (Fig. 2.25).

In Fig. 2.26 the effect of recirculation is seen to be dependent upon the recirculation parameter, the Thring–Newby number, eqns. (2.11a) and (2.26),

$$\theta' = \frac{m_0 + m_a}{2L(G_0 \pi \rho_a)^{1/2}}$$

Here the ratio of enclosed to free jet combustion length is given as a function of $(1/\theta)$.

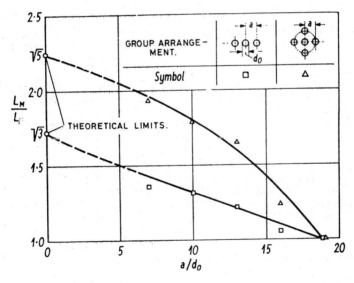

Fig. 2.27. Increased combustion lengths of central jets as a function of the pitch to nozzle diameter ratio for two types of group arrangements (after Eickhoff and Lenze[49]).

2.13 MULTIPLE JET FLAMES

When jet flames are placed in a group arrangement, there can be interference between the individual jet flames resulting in increased length of the flames in the centre of the group. The stability of these flames can also be adversely affected. Allen[50] has shown that, by reducing the pitch to diameter ratio to below a critical value for a group of flames (three in a row), extinction of the central flame can be obtained. Figure 2.27 represents increased combustion lengths of central jets experimentally determined by Eickhoff and Lenze[49] for two types of group arrangements and plotted as a function of the pitch to diameter ratio (a/d_0). They found that when this ratio was increased to beyond $(a/d) \geq 18$ there was no noticeable interaction between the individual jets in the groups.

REFERENCES

1. van der Hegge Zijnen, B. G., 'Measurements of the velocity distribution in a plane turbulent jet of air', *Appl. Sci. Res.* 1958, **A7**, pp. 256–76.
2. Hinze, J. O. and van der Hegge Zijnen, B. G., 'Transfer of heat and matter in the turbulent mixing zone of an axially symmetric jet', *Appl. Sci. Res.* 1949, **A1**, p. 435.
3. Hinze, J. O., *Turbulence*, McGraw-Hill, New York, 1959.
4. Hawthorne, W. R., Weddell, D. S. and Hottel, H. C., 'Mixing and combustion in turbulent gas jets', Third Symposium on Combustion, Flames and Explosion Phenomena, pp. 266–88, Williams and Wilkins, Baltimore, 1951.
5. Thring, M. W. and Newby, M. P., 'Combustion length of enclosed turbulent jet flames', Fourth Symposium on Combustion, pp. 789–96, Williams and Wilkins, Baltimore, 1953.
6. Sunavala, P. D., Hulse, C. and Thring, M. W., 'Mixing and combustion in free and enclosed turbulent jet diffusion flames', *Combustion and Flame* 1957, **1**, 2, pp. 179–93.
7. Ricou, F. P. and Spalding, D. B., 'Measurements of entrainment by axisymmetrical turbulent jets', *J. Fluid Mech.* 1961, **II**, 1, pp. 21–32.
8. Chigier, N. A. and Beér, J. M., 'The flow region near the nozzle in double concentric jets', *Trans. ASME* **86D**, *J. Basic Eng.* 1964, **4**, pp. 797–804.
9. Carmody, T., 'Establishment of the wake behind a disc', *Trans. ASME*, **86D**, *J. Basic Eng.* 1964, **4**, pp. 869–82.
10. Chedaille, J., Leuckel, W. and Chesters, A. K., 'Aerodynamic studies carried out on turbulent jets by the International Flame Research Foundation', *J. Inst. Fuel* 1966, **39**, pp. 506–20.
11. Chigier, N. A. and Gilbert, J. L., 'Recirculation eddies in the wake of flameholders', *J. Inst. Fuel* 1968, **41**, pp. 105–12.

12. Winterfeld, G., 'On processes of turbulent exchange behind flameholders', Tenth Symposium on Combustion, pp. 1265–75, The Combustion Institute, 1965.
13. Miller, R. D. and Comings, E. W., 'Force-momentum fields in a dual jet flow', *J. Fluid Mech.* 1960, **7**, 2, pp. 237–56.
14. Squire, H. B. and Trouncer, J., 'Round Jets in a General Stream', Aeronautical Research Council, Rep. No. 1974, 1944.
15. Corrsin, S., 'Investigation of Flow in an Axially Symmetrical Heated Jet of Air', NACA Wartime Report ARC No. 3L 23-W-94, 1943.
16. Keagy, W. R. and Weller, A. E., 'A study of freely expanding inhomogeneous jets', Heat Transfer and Fluid Mech. Inst., University of California, Berkeley, June 1949, Am. Soc. Mech. Eng., pp. 89–98.
17. Squire, A. B., 'Jet flow and its effect on aircraft', *Aircraft Eng.* 1950, **22**, No. 253.
18. Reynolds, A. J., 'Observations of a liquid-into-liquid jet *J. Fluid Mech.* 1962, **14**, pp. 552–56.
19. Chigier, N. A. and Chervinsky, A., 'Experimental investigation of swirling vortex motion in jets', *Trans. ASME* **34**, E, 2, *J. App. Mech.* 1967, pp. 443–5.
20. Trüpel, T., 'Über die Einwirkung eines Luftstrahles auf die umgebende Luft', *Z. Ges. Turbinenwesen* 1915, pp. 5–6.
21. Abramovich, G. N., *The Theory of Turbulent Jets*, Massachussetts Institute of Technology Press, 1963.
22. Miller, R. D. and Comings, E. W., 'Static pressure distribution in the free turbulent jet', *J. Fluid Mech.* 1957, **3**, pp. 1–16.
23. Prandtl, L., *Essentials of Fluid Dynamics*, Blackie and Son, London, 1954.
24. Ricou, F. P., 'Static pressure measurements in round free turbulent jet', Private Communication, Grenoble, 1963.
25. Schlichting, H., *Boundary-layer Theory*, McGraw-Hill, New York, 1965.
26. Pao, R. H. F., *Fluid Dynamics*, Charles E. Merrill Books, Inc., Columbus, Ohio, 1967.
27. Spalding, D. B. and Patankar, S. V., *Heat and Mass Transfer in Boundary Layers*, Morgan-Grampian Books Ltd, London, 1967.
28. Spalding, D. B. and Wolfshtein, M., Imperial College, London, 1968.
29. Forstall, W. and Shapiro, A. H., 'Momentum and mass transfer in coaxial gas jets', *J. Appl. Mech., Trans. ASME* 1950, **72**, pp. 399–408. Discussion of paper, *J. Appl. Mech., Trans. ASME* 1951, **73**, pp. 219–220.
30. Alpinieri, L. J., 'Turbulent mixing of co-axial jets', *AIAA J.* 1964, **2**, 9, pp. 1560–7.
31. Zhestkov, Glazkov and Gusev, 'Velocity and temperature fields in mixing zone of two plane parallel turbulent jets'. Abramovich, *The Theory of Turbulent Jets*, p. 28, MIT Press, Cambridge, Mass., 1963.
32. Landis, F. and Shapiro, A. H., 'The turbulent mixing of co-axial gas jets', *Heat Transfer Fluid Mech. Inst. (Berkeley)* 1951, pp. 133–46.
33. Corrsin, S. and Uberoi, M. S., 'Further Experiments on Flow and Heat Transfer in a Heated Turbulent Air Jet', NACA, T.N., 1865, April 1949.
34. Libby, P. A., 'Theoretical analysis of turbulent mixing of reactive gases with application to supersonic combustion of hydrogen', *ARS J.* 1962, **32**, pp. 388–96.

35. Vulis, L. A. and Leont'yeva, T. A., 'Co-flowing and counter flowing turbulent jets', *Izv. Akad. Nauk. Kaz. SSR, Ser. Energ.* 1955 **9**; see Abramovich, *The Theory of Turbulent Jets*, p. 36, MIT Press, Cambridge, Mass., 1963.
36. Förthmann, E., Über turbulente Strahlausbreitung, Ingr., — Archiv., 1934, **1.**
37. Weinstein, A. S., Osterle, J. F. and Forstall, W., 'Momentum diffusion from slot jet into moving secondary flow', *J. Appl. Mech.* 1956, **23**, 9, p. 437.
38. Al-Anbari, H., Ph.D. Thesis, The University of Sheffield, 1969.
39. Patrick, M. A., *Sheffield Univ. Fuel Soc. J.* 1965, **16**, pp. 46–61.
40. Siddall, R. G. and Al-Anbari, H., Private communication.
41. Craya, A. and Curtet, R., *Compt. Rend.* 1955, **241**, 1, pp. 621–2.
42. Barchilon, M. and Curtet, R., *J. Basic Eng.* (*Trans. ASME*) 1964, p. 777.
43. Field, M. A., Gill, D. W., Morgan, B. B. and Hawksley, P. G. W., 'Combustion of Pulverised Coal', BCURA, 1967.
44. Curtet, R., *Combustion and Flame* 1958, **2**, pp. 383–411.
45. Burke, S. P. and Schumann, T. E. W., First Symposium on Combustion, p. 2, The Combustion Institute, 1928.
46. Hottel, H. C. and Hawthorne, W. R., Third Symposium on Combustion, p. 254, Williams and Wilkins, 1949.
47. Guenther, R., *Gaswärme* 1966, **15**, p. 376.
48. Kremer, H., 'The spread of nonhomogeneous turbulent free jets and of turbulent diffusion flames', Dissertation, T.H. Karlsruhe, Germany, 1964.
49. Eickhoff, H. and Lenze, B., 'Basic forms of jet flames', *Chem. Ing. Techn.* 1969, **41**, pp. 1095–9.
50. Allen, R. A., Ph.D. Thesis, The University of Sheffield, 1970.

CHAPTER 3

Flame Stabilisation

NOMENCLATURE

a	number of moles per unit volume.
b	boundary of burned gas.
b.r.	blockage ratio $= (d/D)^2$.
c_p	specific heat.
C	concentration.
d	diameter of disc.
D	diameter of orifice.
E	activation energy.
f	molar feed rate.
g	velocity gradient.
h	half height of V-gutter.
ΔH	heat of reaction.
k	thermal conductivity; reaction rate constant.
K	Karlovitz number; stretch factor.
K_i	rate of change of n_i due to reaction.
l	scale of turbulence; eddy diameter.
L	length.
$\dot{m}$	rate of mass transfer.
m_i	molecular weight.
M	mass flux; mass flow rate.
n	number of moles.
p	pressure.
$\dot{Q}$	heat generated or removed per unit volume.
r	radial co-ordinate.
R	universal gas constant; pipe radius.
S	flow velocity; flame velocity; surface of reverse flow zone.
t	time.
T	temperature.
$\bar{u}$	mean axial velocity.
u'	velocity fluctuation axial.

$\overline{u'v'}$ shear stress.
U velocity.
v volume flow rate.
v' radial velocity fluctuation.
V reactor volume.
V_i diffusion velocity.
w' tangential velocity fluctuation.
x length co-ordinate.
X volume of fuel gas in mixture-fraction of stoichiometric.
y co-ordinate.
Z frequency factor.

Greek Symbols
α angle; forebody divergent angle; mixture ratio.
δ flame front thickness.
η co-ordinate.
η_0 width of preheat zone.
μ dynamic viscosity.
ξ co-ordinate.
ρ density.
τ time.
φ equivalence ratio.
ψ stream function.

3.1   FLAME PROPAGATION

In burner flames the flame is propagating against the flow of the reagents
and its position is stationary. Variation in input conditions such as fuel
flow rate, air/fuel or preheat can cause these flames to become non-
stationary or unstable. A burner flame is considered to be stable over a
range of an input parameter if variation of such a parameter within this
range does not cause the flame to blow off or to flashback into the burner
tube.

One of the basic concepts in flame theory is that of flame propagation.
This refers to the propagation of the zone of burning or of the combustion wave through a combustible mixture. It is generally appreciated that
the ignition source is a source of heat. It may, however, also produce
atoms and free radicals which may act as chain carriers in the chemical

reaction. Once the heat flow and the diffusion of these active species have initiated chemical reaction in the adjacent layer of the combustible medium, this layer itself becomes the source of heat and of chain carriers and is capable of initiating reaction in the next layer. A quantitative theory of flame propagation will therefore have to be based on the transfer of heat and mass from the reaction zone to the unburned mixture.

3.2 ONE-DIMENSIONAL LAMINAR COMBUSTION WAVE

Most of the discussions on flame propagation have been concerned with the case of the adiabatic plane laminar combustion wave.[1] Although such discussions cannot directly be applied to practical systems of combustion, it is possible to make use of the concepts and results for predicting performance and stability in a volume element in which the reactants are premixed and the air/fuel ratio of the unburned state is uniform throughout the volume. In this case the combustion problem can be illustrated by a single step reaction. The propagation is assumed to be 'thermal', *i.e.* mainly by the transport of heat from the reaction zone to the unburnt gas.

Temperature and concentration profiles in a plane adiabatic laminar wave are shown in Fig. 3.1. Heat flows from the boundary *b* of the burned gas towards the boundary of the unburned mixture. In the preheat zone

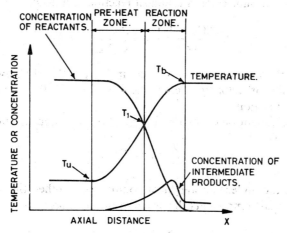

Fig. 3.1. *Temperature and concentration distributions in an adiabatic plane combustion wave.*

a mass element gains more heat by conduction from downstream than it loses to neighbouring elements upstream—the temperature curve is convex towards the abscissa. At the inflexion point where the reaction zone starts it loses more heat to upstream elements than it gains, but its temperature goes on rising because of the heat of the exothermic reaction. At the fully burned boundary the concentration of the reactants is exhausted and the temperature curve levels off. Corresponding to the shape of the temperature profile, the gradient of the heat flux,

$$\frac{d}{dx}\left(k\,\frac{dT}{dx}\right)$$

will have a positive sign in the preheat zone and a negative sign in the reaction zone. The thermal expansion due to rapid rise of the temperature takes place perpendicular to the direction of flow.

The reactant concentration changes in a similar manner in the wave since the loss of reactants from a mass element is symmetrically opposite to the gain in heat. In the preheat zone, concentration decays by diffusion only, while in the reaction zone there is a rapid depletion by chemical reaction.

Under steady state conditions, the number of moles of any component of the gas in a volume element is constant with time. Molecules of any component are transported into and out of the volume element by convection and diffusion. The only process that can cause the formation or disappearance of molecules is chemical reaction and hence

$$\frac{dn_i(S + V_i)}{dx} = \left(\frac{\partial n_i}{\partial t}\right)_{\text{chem.}} = K_i \tag{3.1}$$

where n_i is the number of moles of component i, S is the flow velocity, V_i is the diffusion velocity of component i, and K_i is the rate of change of n_i due to reaction.

New mass cannot be created by chemical reaction and therefore

$$\sum m_i K_i = 0 \tag{3.2}$$

where m_i is the molecular weight of component i. When the above sum is taken over all components this leads to the statement

$$\frac{d(\rho S)}{dx} = 0 \tag{3.3}$$

which integrates to

$$\rho S = \rho_u S_u = M \tag{3.4}$$

where M is the mass flow rate per unit area.

The energy equation for a unit volume of the wave can be written as

$$\frac{d}{dx}\left(k\frac{dT}{dx}\right) - c_p\rho_u S_u\frac{dT}{dx} + \dot{Q} = 0 \tag{3.5}$$

where $\dot{Q}$ is the heat generated per unit volume and unit time.

The equations of chemical reaction, conservation of energy together with the energy equation, can be solved for the mass flow and hence for the wave propagation velocity after introducing appropriate boundary conditions. At the hot boundary b, the gradients of temperature and of concentration vanish together with the rate of the chemical reaction. At the cold boundary u, an artificial condition has to be introduced in the form of an 'ignition temperature' or a 'flame holder'.[2] The solution of the differential equations gives the mass flow as an 'eigenvalue' and the temperature and concentration distributions within the wave can then be determined in detail.

Zeldovich and Frank-Kamenetsky[3] give an approximate solution for the energy eqn. (3.5) as

$$S_u = \frac{\left(2k\int_{T_u}^{T_b}\dot{Q}\,dT\right)^{1/2}}{\rho_u c_p(T_b - T_u)} \tag{3.6}$$

where S_u is the velocity of steady propagation referred to the unburned state. Further details of the solution of the flame equations are given by Friedman and Burke.[4] As the reaction proceeds in a volume element, the rate of reaction and thus the heat generated will vary as a function of the 'reactedness' of the mixture. At the beginning of the process the reactant concentration is high but the temperature is low. As the temperature increases a maximum rate is obtained beyond which the rate drops again due to the exhaustion of the reactant concentration. This can be illustrated by a curve giving the volumetric heat release in a closed system as a function of reactedness, or alternatively as a function of the temperature of the mixture (Fig. 3.2).

It follows from the strong temperature dependence of the rate equation that changes in air/fuel ratio also have an effect upon the reaction rate curve (Fig. 3.2) mainly by the way in which the theoretical temperature increases as the air/fuel ratio approaches the stoichiometric. Changes in reaction rate due to variation in initial temperature show a similar trend for the same reason.

There is not sufficient information available on the kinetics of the chemical reaction for $\dot{Q}_{max}$ to be predicted. Knowing, however, the

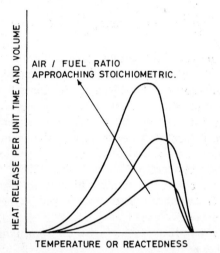

Fig. 3.2. *Heat release rate in a closed system as a function of temperature or reactedness.*

general shape of the heat release rate–temperature curve and by determining the mean value of the volumetric heat release rate, it is possible to estimate $\dot{Q}_{max}$ with good approximation.

Spalding[6] suggested rewriting eqn. (3.6) as

$$\frac{\int_{T_u}^{T_b} \dot{Q}\,\mathrm{d}T}{T_b - T_u} = \dot{Q}_{\text{mean}} = \frac{c_p{}^2 S_u{}^2 \rho_u{}^2 (T_b - T_u)}{2k} \tag{3.7}$$

Thus the mean value of the volumetric rate of reaction can be determined from easily measurable quantities and a general idea of the form of the reaction rate curve can then yield the value of $\dot{Q}_{max}$.

3.3   PREMIXED LAMINAR FLAME HOLDING

The condition for flame stabilisation in a flow field of non-uniform velocity is that there is a point in the flow field where the flow velocity is equal and opposite to the velocity of the combustion wave. At all other points the flow velocity of the combustible mixture exceeds the velocity of the combustion wave.

In the case of burner flames where the flow velocity exceeds the burning velocity everywhere in the field, the combustion wave is driven back causing *blow off*. Conversely, where the burning velocity exceeds the flow velocity, the combustion wave propagates into the burner tube, a phenomenon called *flashback*. It is significant that, in practical systems such as a bunsen flame, it is not only the velocity field that is not uniform, but the burning velocity will vary also as a function of space. The burning velocity depends upon a number of variables such as the mixture strength, the theoretical flame temperature and material constants such as the thermal and molecular diffusivities of the mixture. In practical systems, heat and active species produced in the reaction zone can be removed by sink effects or by dilution. When the combustion wave propagation is divergent, the enthalpy transferred by conduction from the reaction zone is not fully returned by convection to the element of the combustion wave from which it originated. It is therefore necessary to consider the spatial variations of both the flow velocity and of the burning velocity for flame stability discussions.

Lewis and von Elbe[2] have considered the stability of a laminar pre-mixed jet flame. Under conditions of stability the combustion wave is 'anchored' at some position downstream from the burner port. The boundary layer formed on the inner wall of the burner extends beyond

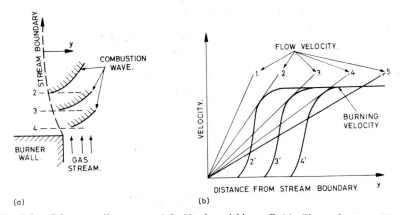

(a)   (b)

Fig. 3.3.   *Schematic illustration of flashback and blow off.* (a) *Flame front positions above burner rim for various flow velocities.* (b) *Burning velocity and flow velocity close to the burner rim.*

the burner exit. The flame is stabilised within the boundary layer region where velocities are lower than in the central region of the flow. The rim, however, is also a sink for heat and for active species. This in turn adversely affects flame stabilisation very close to the rim.

In a very thin layer close to the burner wall the radial distribution of the velocity can be assumed to be linear. Velocity profiles corresponding to a range of flow rates are illustrated by the straight lines 1–5 in Fig. 3.3. The curves 2′, 3′ and 4′ represent lateral distributions of the burning velocity for arbitrary flame positions 2, 3 and 4. The reduction in burning velocity due to the sink effect of the burner rim is greatest when the position of the combustion wave is closest to the rim. Two extreme cases are represented by the velocity gradients 1 and 5. When the flow rate is very high, such as that illustrated by gradient 1, the flow velocity of the mixture exceeds the burning velocity everywhere and the flame 'blows off'. When, on the other hand, the flow rate of the mixture is so low that the local burning velocity exceeds at some point the flow velocity, 'flashback' will occur.

Between the flow rates corresponding to the gradients 2 and 4, which are limits for blow off and flashback respectively, the flame will be stable.

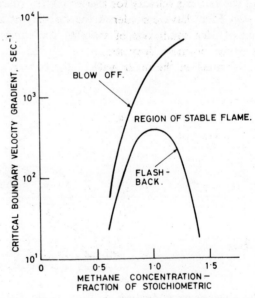

Fig. 3.4. *Critical boundary velocity gradients for flashback and blow off in cylindrical tubes. Methane–air mixtures at room temperatures and atmospheric pressure.*

This is because of the increase in burning velocity near the stream boundary as the position of the combustion wave moves downstream from the rim. If the flame is originally stabilised in position 4 and the flow rate is increased to produce the gradient 3, the flame will move downstream, but will now stabilise in position 3 because of the increase in burning velocity. This will happen again as the gradient is increased to its limiting value 2. Beyond this position, however, there is no further increase in burning velocity that could compensate for a further increase of the velocity gradient. This is because the dilution of the flame gases by the surrounding gas becomes effective at some distance from the burner rim.

This rather simple representation of blow off and flashback phenomena indicates that these phenomena should correlate well with the velocity gradient in the region close to the burner rim.

It has been shown by Lewis and von Elbe[2] that there is good experimental correlation between the boundary velocity gradients at blow off and flashback and the mixture ratio of a gas–air mixture (Fig. 3.4).

Assuming fully developed laminar pipe flow, the boundary velocity gradient can be calculated from the Poiseuille equation

$$U = n(R^2 - r^2) \tag{3.8}$$

where U is the velocity at radius r, R is the pipe radius, r is the radial co-ordinate, and $n = \Delta p/4\mu L$, where Δp = pressure difference over length L, and μ = dynamic viscosity.

Differentiating eqn. (3.8) and setting $r = R$, the velocity gradient g can be given as

$$g = -2nR \tag{3.9}$$

The boundary velocity gradient for a given pipe diameter can also be obtained from the volume flow rate. In a cylindrical tube the volume flow rate can be given as

$$v = 2\pi \int_0^R Ur \, dr$$

Substituting $U = n(R^2 - r^2)$ and integrating we have

$$v = \frac{\pi}{2} nR^4 \tag{3.10}$$

and by combining eqns. (3.9) and (3.10) we have

$$g = \frac{-4}{\pi} \frac{v}{R^3} \tag{3.11}$$

Equation (3.11) gives the boundary velocity gradient as a simple function of volume flow rate and radius of the tube.

Additional information on flame stabilisation is given by Wohl et al.[7] and by Williams et al.[8]

More recently, Reed[27] has challenged the classical theory of Lewis and von Elbe on the stabilisation of aerated flames on a burner rim. He suggests that blow off may result from a reduction in the reaction rate caused by the enthalpy loss from the stabilising region due to shear flow, rather than from the fact that the gas velocity exceeds the local burning

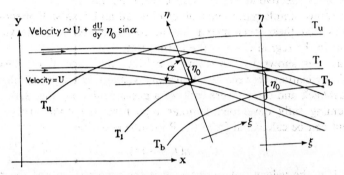

Fig. 3.5.   Schematic diagram of divergent flame propagation.

velocity throughout the whole flow field. The case of divergent propagation of a combustion wave in which the flame propagates in a flow field with appreciable velocity gradients resulting in flame extinction has been treated by Karlovitz et al.[26]

A schematic diagram of divergent flame propagation is shown in Fig. 3.5. The T_u and T_b curves are isotherms and mark the boundaries of the unburned and fully burned gases respectively. The T_1 isotherm passes through the inflexion points of the temperature curves (see Fig. 3.1) and thus separates the preheat and reaction zones. At the intersection of a stream-line with the T_1 surface, the mass flow normal to the T_1 surface can be written as

$$\rho U \cos \alpha \qquad (3.12)$$

Because of the velocity gradient in this flow field, the velocity at a distance η_0 from the previously considered point is higher and can be given as

$$U + \left(\frac{dU}{dy}\right) \eta_0 \sin \alpha \qquad (3.13)$$

The mass flow rate normal to the combustion wave at distance η_0 from the T_1 isotherm is then

$$\rho \left[U + \left(\frac{dU}{dy}\right) \eta_0 \sin \alpha \right] \cos \alpha \qquad (3.14)$$

The increase in area of a combustion wave element as it travels the distance η_0 can then be given as the ratio of mass flows through the element over the distance of η_0. This is

$$1 + \left(\frac{dU}{dy}\right)\left(\frac{\eta_0}{U}\right) \sin \alpha \qquad (3.15)$$

Close to blow off conditions, $\sin \alpha$ is close to unity so that the second term may be written as

$$\frac{dU}{dy} \frac{\eta_0}{U} = K \qquad (3.16)$$

This term, the 'stretch' factor or Karlovitz number as it is called by Lewis and von Elbe[2], can be considered as the dimensionless criterion for the increase in area that the combustion wave surface undergoes in a flow field with velocity gradients.

It can be shown[26] that η_0 in the above equation is a measure of the width of the preheat zone of the wave and can be given as a function of measurable data as

$$\eta_0 = \frac{k}{c_p \rho_u S_u} \qquad (3.17)$$

The measure of stretch therefore depends on the magnitude of the velocity gradient relative to the width of the preheat zone.

Reed[27] has extended the application of the flame stretch theory to the case of aerated burner flames and has shown that the correlation between mixture composition and the boundary velocity gradient at blow off can be explained by the flame stretch theory.

The boundary velocity gradient theory of blow off can thus be considered as a special case of the more general flame stretch theory of blow off.

Correlating a large number of experimental blow off data with the Karlovitz criterion, Reed recommends the following expression:

$$g_b \simeq (0.23 \rho c_p S_u^2 / k)[1 - (1 - X^{6.4})\alpha] \qquad (3.18)$$

where g_b is the boundary velocity gradient at blow off, S_u is the laminar flame speed, k is the thermal conductivity of the gas, X is the volume of fuel gas in the mixture expressed as a fraction of stoichiometric, and α is a constant, $\alpha = 0$ for premixed flames and $\alpha = 1$ for flames with secondary combustion.

The above equation is valid for $\alpha = 0$ or $\alpha = 1$ and $X < 1.36$.

Further support for the flame stretch theory of blow off was given by Edmondson and Heap[28] on the basis of their studies of inverted methane–air flames.

3.4   TURBULENT COMBUSTION

The first published results of measurements on turbulent combustion were made by Damköhler[9] in an experimental study with propane/air mixtures on burners of very small diameters between 0·13 and 0·27 cm. Damköhler attempted to measure the effect of turbulence on flame propagation, and he put forward a theoretical concept for the interpretation of the experimental results. The experiments showed that the turbulent burning velocity is a function of the velocity of flow. In Fig. 3.6 Damköhler's results are plotted in the form of the ratio of the turbulent to laminar burning velocity as a function of the flow velocity of a combustible mixture. It can be seen that the ratio of the turbulent to laminar burning velocity rises rapidly at first and then levels off with an increase in Reynolds number. Since the laminar burning velocity for a given mixture concentration is constant, the curve represents the variation in turbulent flame

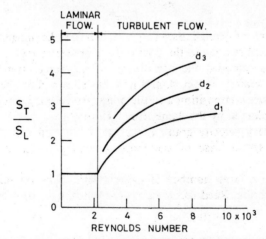

Fig. 3.6.   *Ratio of turbulent to laminar burning velocity as a function of Reynolds numbers and burner tube diameter (Damköhler[9]).*

velocity as a function of the Reynolds number. In another set of experiments, Williams and Bollinger[10] have used acetylene, ethylene and propane, with a variation in burner diameter from 6 to 30 mm. Their results, in which turbulent and laminar burning velocities for these hydrocarbon fuels are compared, are represented in Fig. 3.7.

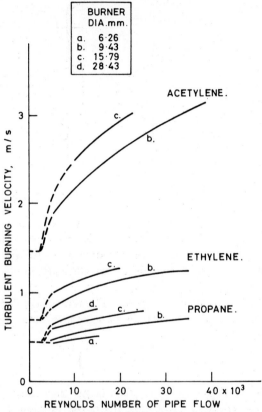

Fig. 3.7. *Variation of turbulent burning velocity with Reynolds number of pipe flow (Williams and Bollinger[10]).*

Damköhler and Shchelkin[11] proposed two theoretical models for the effect of turbulence on flame propagation, but neither of these has as yet been satisfactorily proved experimentally. In the first model, the assumption is made that the scale of turbulence is very small in comparison with the flame front thickness, $l_1 \ll \delta$. In this case, the effect of turbulence is to increase the rate of transfer of heat and of active species through the

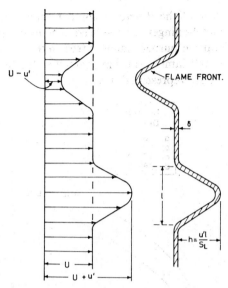

Fig. 3.8.   Flame front in turbulent stream with large scale turbulence.

pre-ignition and reaction zones. It can be shown from simple dimensional analysis that the laminar burning velocity is proportional to the square root of the thermal diffusivity. The introduction of turbulence increases this molecular transfer coefficient by an amount equal to the coefficient of eddy diffusivity, *i.e.* by the product of $u'l_1$, where u' is the r.m.s. fluctuation velocity and l_1 is the scale of turbulence. The ratio of the flame velocity in the turbulent stream S_T to laminar flame velocity S_L may then be written as

$$\frac{S_T}{S_L} = \left(\frac{(k/c_p\rho) + u'l_1}{(k/c_p\rho)} \right)^{1/2} \tag{3.19}$$

When the scale of turbulence is much larger than the thickness of the flame front, the effect of turbulence upon the burning velocity of the mixture is due to the indentations in the flame front caused by the fluctuating velocity. Figure 3.8 represents a model of the flame front in a turbulent stream with large scale turbulence. Assuming that the indentations are conical and that the entire combustion wave is distorted into cones, Shchelkin assumed that the ratio of the turbulent to laminar burning velocities, S_T/S_L, equals the ratio of the average cone area to the area of the average cone base. The base area is taken as proportional to l^2, where l is the eddy diameter, and the height of the cone proportional to u', while the time during which an element of the combustion wave is associated with an eddy moving in the direction normal to the wave is $t = l/S_L$.

From simple geometrical considerations, the cone area equals the cone base times $(1 + 4h^2/l^2)^{1/2}$, where h is the height of the cone and l is the base diameter.

Since $h = u'l/S_L$,

$$S_T = S_L \left[1 + \left(\frac{2u'}{S_L}\right)^2\right]^{1/2} \tag{3.20}$$

which reduces to $S_T \sim u'$ for large values of $(u'/S_L)^2$.

The actual conditions in a turbulent flame do not strictly agree with either of the assumptions that the scale of turbulence is much smaller or much larger than the thickness of the combustion zone in the flame. Eddy sizes will be distributed over a wide spectrum, with the small eddies satisfying eqn. (3.19) and large eddies eqn. (3.20).

Scurlock[12] recommends a generalised equation which reduces to eqn. (3.19) when l_2/δ is small and to eqn. (3.20) for the case when l_2/δ is large (l_1 and l_2 are the Lagrangian and Eulerian scales of turbulence respectively)

$$\frac{S_T}{S_L} = \left\{\left[1 + \left(\frac{2u'}{S_L}\right)^2 \left(\frac{l_2/\delta}{K_1 + (l_2/\delta)}\right)\right] \right.$$
$$\left. \times \left[1 + \left(\frac{cu'l_2}{k/c_p\rho}\right)\left(\frac{1}{1 + (l_2/K_2\delta)}\right)\right]\right\}^{1/2} \tag{3.21}$$

where K_1 and K_2 are constants and cl_2 has been substituted for l_1.

Tables 3.1 and 3.2 contain summaries of turbulent flame theories and of experimental data on turbulent flames as presented by Lefebvre and Reid.[13]

The main conclusions can be given as follows:

(a) the turbulent burning velocity S_T is independent of the scale of turbulence.

(b) experimental results in general support the wrinkled flame theory. At low velocity of flow, S_T is determined by the laminar flame speed and the intensity of turbulence. It follows that at low speed S_T will depend also on the air/fuel ratio (because S_L is dependent upon the mixture ratio).

(c) at high intensities of turbulence or at high inlet flow velocities the turbulent flame speed S_T is roughly proportional to the intensity of turbulence and to the inlet velocity, but is independent of the air/fuel ratio.

THE INFLUENCE OF TURBULENCE ON STRUCTURE AND PROPAGATION OF ENCLOSED FLAMES

TABLE 3.1

Turbulent Flame Theories
(after Lefebvre and Reid[13])

Investigator	Postulated structure	Equations	Main conclusions
Damköhler[9]	Wrinkled laminar flame	$S_T = S_L + u'$ At high velocities this approaches $S_T = u'$	S_T is independent of scale of turbulence. At low velocities S_T is determined by laminar flame speed and turbulent velocity. At high velocities S_T is determined solely by turbulent velocity, *i.e.* S_T is independent of fuel/air ratio
Shchelkin[11]	Wrinkled laminar flame	$S_T = S_L \left[1 + B \left(\dfrac{u'}{S_L} \right)^2 \right]^{0.5}$ At high velocities this approaches $S_T = u'$	Broadly in agreement with Damköhler
Karlovitz et al.[14]	As above but with augmentation by flame-generated turbulence	For weak turbulence $S_T = S_L + u'$ For strong turbulence $S_T = S_L + (2S_L u')^{0.5}$ where $u' = \dfrac{S_L}{\sqrt{3}} \left(\dfrac{\rho_u}{\rho_b} - 1 \right)$	S_T is independent of scale of turbulence. Laminar flame speed is a more important factor than turbulence even at very high levels of turbulence
Scurlock and Grover[15]	As above but with augmentation by flame-generated turbulence	$S_T = S_L \left[1 + C_3 \left(\dfrac{\bar{y}}{l} \right)^2 \right]^{0.5}$ $\bar{y}$ is dependent on approach stream turbulence and flame-generated turbulence	S_T is dependent on scale of turbulence and on laminar flame speed. For confined flames, in which $U \gg S_T$, approach stream turbulence is outweighed by flame-generated turbulence

TABLE 3.1—*contd.*

Investigator	Postulated structure	Equations	Main conclusions
Summerfield et al.[16]	Distributed reaction zone	None directly applicable	Argues that wrinkled laminar flame description of turbulent flame should be abandoned in favour of a distributed reaction zone model. Offers little convincing physical evidence in support of this view
Spalding[17]	Flame propagation determined by rate of entrainment of cold mixture by hot gases	None directly applicable	Flame spreading dictated by laws of jet entrainment. S_T independent of scale of turbulence and percentage turbulence and proportional to inlet velocity. S_T independent of laminar flame speed except indirectly through relationship between S_L and density ratio ρ_u/ρ_b

TABLE 3.2

Experimental Data on Enclosed Flames
(after Lefebvre and Reid[13])

Investigators	Variables studied	Test conditions	Apparatus	Results
Williams et al.[8]	Percentage turbulence. Turbulence scale. Inlet velocity. Stabiliser size and shape. Fuel/air ratio. Fuel type.	Turbulence: 0·4 to 8·0%. Turbulence scale: 0·01 to 0·08 in. Inlet velocity: 20 to 350 ft/sec. Inlet temperature: 300 to 340K. Pressure: atmospheric. Fuel: city gas and propane	Horizontal duct 17 in long of rectangular cross section 3 in × 1 in. Duct fitted with windows for direct and Schlieren photography. Bluff-body flame holders in form of single and multiple rods, 30° Vee gutters and flat plates	At all velocities flame holder dimensions had negligible effect on flame propagation except at limit mixtures. No effect of turbulence up to 2·3%. Higher turbulence level produced appreciable effect. At low velocity ($U > 50$ ft/sec) S_T varied with fuel/air ratio. At high velocity ($U > 100$ ft/sec) S_T was independent of fuel/air ratio and roughly proportional to inlet velocity
Wohl et al.[18]	Percentage turbulence. Inlet velocity. Stabiliser size.	Turbulence: normally 0·4%. Increased by screens to 9%. Inlet velocity: 24 to 82 ft/sec. Room temperature and pressure. Fuel: stoichiometric propane/air	Horizontal duct 10 in long of rectangular cross section 2 in × 1·5 in. Duct fitted with glass windows for direct and Schlieren photography. Flat plate flame holders of thickness 0·117, 0·247 and 0·478 in	S_T increased with both percentage turbulence and inlet velocity as described by the equation $$\frac{S_T}{S_L} = 1 + 0·262T + 1·40\left(\frac{U}{24}\right)^{1·12}$$ or $$\frac{S_T}{S_L} = 1 + 26·2\frac{u'}{U} + 1·40\left(\frac{U}{24}\right)^{1·12}$$ This form of equation supports wrinkled flame concept

TABLE 3.2—contd.

Investigators	Variables studied	Test conditions	Apparatus	Results
Wright and Zukoski[19]	Inlet velocity. Inlet temperature. Stabiliser size and blockage. Fuel/air ratio. Fuel type.	Very low turbulence. Inlet velocity: up to 440 ft/sec. Inlet temperature: 373 to 520K. Pressure: atmospheric. Fuels: gasoline and hydrogen	Horizontal duct 15 in long of rectangular cross section 6 in × 3 in. Fitted with transparent side walls for direct and Schlieren photography. Cylindrical flame holders of 0·125 to 2·0 in diameter	Rate of flame spreading from a bluff body was found to be 'remarkably independent of approach stream speed, temperature, fuel/air ratio and fuel type, as long as flame is turbulent and flow is everywhere subsonic.* Results strongly indicate that the local flame speed is proportional to flow speed.' *This result is in conflict with the wrinkled flame model and all other theories
Lefebvre and Reid[13]	Percentage turbulence. Inlet velocity. Fuel/air ratio.	Turbulence: 2 to 14%. Inlet velocity: 30 to 250 ft/sec. Inlet temperature: 298K. Pressure: atmospheric. Fuel: propane	Horizontal duct 12 in long of 4 in × 4 in cross section. Transparent side walls for Schlieren photography. Flame stabilisation provided by pilot burner 0·75 in diameter	S_T increased with both percentage turbulence and inlet velocity as described by the equation $$\frac{S_T}{S_L} = 1 + 0\cdot0043T + 0\cdot04U$$ or $$\frac{S_T}{S_L} = 1 + 0\cdot43u' + 0\cdot04U$$ The results fully support the wrinkled laminar flame concept. In particular, S_T and flame spreading rate varied with percentage turbulence and fuel/air ratio at all velocities.

3.5   FLAME STABILISATION BY BLUFF BODIES

When an obstacle is placed in the flow of a combustible mixture, the flow velocity is reduced in a boundary layer near the solid wall and the chances for the flame speed to match the flow velocity at some region in the flow field—a requirement of flame stabilisation—are improved. If the obstacle is a bluff body, *i.e.* non-streamlined, then as the fluid is accelerated a flow velocity is reached where the adverse pressure gradient downstream from the obstacle is strong enough to cause separation of the boundary layer and a recirculating vortex system is set up in the wake of the bluff body. The rate of transfer of mass and of heat between the wake and the flow past the wake is high and, if there is combustion in the flow surrounding the recirculation zone, hot combustion products will penetrate the vortex and be carried upstream where they can mix with the fresh combustible mixture and prepare it for ignition.

Except for the conditions near flame extinction, it can be assumed that the gas composition and temperature of the recirculating flow will be close to those of the fully reacted mixture. (This assumption implies that the wake of a bluff body flame holder is a 'well stirred' zone.) Apart from transferring chemically active species from the combustion products into the fresh mixture, the wake also acts as a reservoir of heat. The transfer of heat and of active species to the combustible mixture, together with the retardation of the flow velocity in the wall boundary layer of the bluff body, enables flames to be stabilised at approach velocities far exceeding the flame propagation velocity.

A discussion on the mechanism of bluff body flame stabilisation can conveniently be grouped as

   (i) the propagation of flame in regions of flow with high velocity gradients, and

  (ii) the fluid dynamic characteristics of the wake, such as its size, and the recirculating mass flow rate as a function of the dimensions of the obstacle and of the mass flow rate and velocity of the approach flow. These characteristics ought to be determined for conditions of both isothermal flow and with flame.

Zukoski and Marble[20] have shown from their experimental studies with gasoline–air mixtures, various flame holder geometries and stream velocities corresponding to $10^3 < \mathrm{Re} < 10^5$ that blow off limits can be determined by the time which a mass element of the fresh mixture takes to sweep past the recirculation zone.

This can be given as

$$\tau = \frac{L}{U} \tag{3.22}$$

where L is the length of the wake in metres and U is the flow velocity in m/sec.

It is a requirement of flame stabilisation that the time τ be longer than the time necessary for preparing the fresh mixture for ignition.

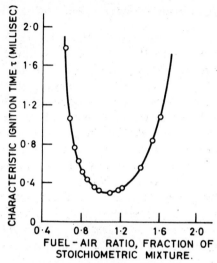

Fig. 3.9. *Characteristic ignition time as a function of mixture composition (Zukoski and Marble[20]).*

Characteristic ignition times averaged in the above experiments were found to be about 3×10^{-4} sec and independent of flow conditions. The ignition times were, however, strongly dependent on mixture composition (Fig. 3.9).

Lewis and von Elbe[2] made use of experimental data obtained by Zukoski and Marble[20] and applied the flame stretch concept to predict the characteristic time during which a mass element of the mixture passes the wake. As can be seen in Fig. 3.10, it takes the combustion wave anchored in a stream tube near the solid boundary the same time to propagate along the path y at a velocity S_u as it does for a mass element of fresh mixture to traverse path L at a velocity U. Therefore

$$\tau = \frac{L}{U} = \frac{y}{S_u} \tag{3.23}$$

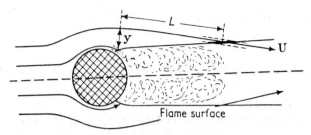

Fig. 3.10.   *Flame propagation in the wake of a cylindrical bluff body (Lewis and von Elbe[2]).*

Along the path y the flow velocity U increases from its value at the solid boundary S_u, determined by the value of the gradient dU/dy, and the wave is therefore stretched. This means that heat which is diffusing out of the reaction zone from an element of the wave is not fully returned by convection to the same element as it moves along the wake. Because of the 'heat reservoir' nature of the recirculation zone, however, the heat lost by flame stretch can be replenished from the heat stored within the wake as long as the wave element in question is sufficiently close to the wake. Because of this, the flame will not blow off. By making the assumption that the stretch factor $K = (\eta_0/U)(dU/dy)$ (where η_0 is the width of the preheat zone) is everywhere at its critical value along the path y, Lewis and von Elbe[2] obtained an upper limit of y, the distance of flame travel. Assuming the critical value of the Karlovitz number to be approximately equal to unity, they integrated the differential equation

$$\frac{1}{U}\,dU = \frac{1}{\eta_0}\,dy \qquad\qquad (3.24)$$

between the limits S_u and U_{max}.

This yields

$$\log_{10}\frac{U_{max}}{S_u} = 0\cdot434\left(\frac{y}{\eta_0}\right) \qquad\qquad (3.25)$$

After substituting experimentally determined data for $S_u = 0\cdot4$ m/sec (hydrocarbon–air flame) and $U_{max} = 100$ to 200 m/sec, the term $\log_{10}(U_{max}/S_u) = 2\cdot4$ to $2\cdot7$. Taking $2\cdot6$ for the log term, $y = 6\eta_0$ and with $\eta_0 \simeq 5 \times 10^{-5}$ m for stoichiometric hydrocarbon–air mixtures,

$$y \simeq 3 \times 10^{-4}\ \text{m}$$

and

$$\tau \simeq \frac{y}{S_u} \simeq 7\cdot5 \times 10^{-4}\ \text{sec}$$

This value compares favourably with that determined experimentally by Zukoski and Marble[20]: $\tau_{exp} = 3 \times 10^{-4}$ sec.

Another method of determining the conditions of flame extinction is to consider the thermal balance of the torroidal vortex region of the wake of a bluff body. In this analysis, the assumption is made that the vortex region of the wake is 'well stirred', *i.e.* the concentration of gas constituents and the temperature of the combustion products are uniform throughout

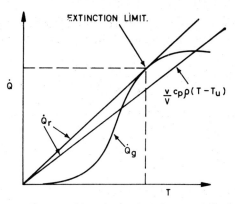

Fig. 3.11.   *Heat release and heat removal curves in a well-stirred reactor.*

the recirculation zone. Curves of specific heat release rates as a function of temperature or 'reactedness' will now be of sigmoid shape, *i.e.* open-ended as shown in Fig. 3.11. A simple heat balance shows the way in which important parameters such as air–fuel ratio, initial temperature of the mixture and the rate of supply of fresh reagents affect the stability of a continuous flow system. The significance of this is that recirculation zones in combustors can generally be considered as continuous flow systems with uniform temperature and concentration.

Assuming a well stirred zone, we can write for the molal feed rate of substance A

$$f_a = v \times a + Vka \tag{3.26}$$

where v is the volume flow rate, a is the number of moles per unit volume, V is the reactor volume and k is a reaction rate constant.

This may be rewritten as

$$a = f_a/(v + Vk) = f_a/v(1 + k\bar{t}) \tag{3.27}$$

where $\bar{t} = V/v$, the mean residence time.

The rate of heat generation per unit volume is

$$\dot{Q}_g = -ka\,\Delta H \tag{3.28}$$

where ΔH is the heat of reaction per mole of species A.

By substituting for a from eqn. (3.27)

$$\dot{Q}_g = -\frac{kif_a\,\Delta H}{(1 + ki)V} \tag{3.29}$$

By further substituting $k = Z \exp(-E/RT)$, where Z is the frequency factor, E the activation energy and R the gas constant, we have

$$\dot{Q}_g = \frac{-Zf_a i\,\Delta H}{[\exp(E/RT) + Zi]V} \tag{3.30}$$

The rate of heat removal, per unit volume, by convection from the well stirred region is

$$\dot{Q}_r = \frac{v}{V}\rho c_p(T - T_u) \tag{3.31}$$

The intersect of the heat release curve $\dot{Q}_g$ and the heat removal curve $\dot{Q}_r$ (Fig. 3.11) gives the solution of the heat balance equation for steady state conditions, *i.e.*

$$\rho\,\frac{v}{V}c_p(T - T_u) = \dot{Q}_g \tag{3.32}$$

and the slope of the heat removal line is $(v/V)c_p\rho$.

At extinction, the point of tangency of the heat removal curve and of the heat release curve is at a value $\dot{Q}$ which is near to $\dot{Q}_{max}$, so that we can write

$$\frac{v_{ext}}{V}c_p\rho(T_b - T_u) = \dot{Q}_{max} \tag{3.33}$$

Increasing flow rate through the volume is represented by increasing slope of the heat removal curve. At the extinction limit the heat removal curve is at a tangent to the heat generation curve and any further increase in flow rate will result in extinction (Fig. 3.11).

Applying the heat release equation to bluff bodies, we can write

$$\dot{Q}_{max} = f(p, T_u, \varphi)$$

If d is the diameter of the bluff body, U_{ext} is the approach flow velocity at extinction and p is the gas pressure, then

$$\dot{m}_{ext} \propto U_{ext} \times \rho \times d^2$$

$$\rho_{gas} \propto p$$

Maximum heat release rate $\dot{Q}_{max} \, V \propto p^n d^3$ ($n =$ order of reaction). At the extinction limit

$$\frac{U_{ext} p d^2}{p^n d^3} = \text{constant}$$

and hence stability requires that

$$\frac{U_{ext}}{p^{n-1} d} = \text{constant}$$

This simple relationship shows that the blow off velocity is proportional to the bluff body diameter and it is also proportional to the pressure (p^{n-1}), where n is the order of the chemical reaction.

3.6 FLOW IN THE WAKE OF BLUFF BODIES

The flow in the wake of a bluff body inserted into a main stream of air is influenced by the size and shape of the bluff body and by the surrounding flow system. The flow system for the wake of a bluff body inserted into a

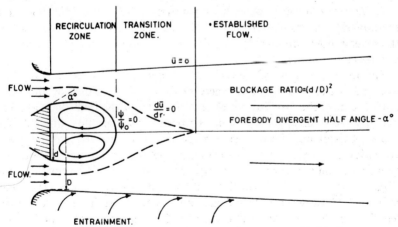

Fig. 3.12. *The flow system for flow in the wake of a bluff body.*

uniform air stream is shown in Fig. 3.12. The air issues from the nozzle in the form of an annular jet which expands as a result of entrainment from the surroundings.

A toroidal vortex is formed within the recirculation zone and is considered to be a closed loop driven by the main stream such that there is no net mass flow across the boundaries. The boundary of the recirculation zone is determined from the radial points at which the forward mass flow equals the reverse flow at that axial station. This boundary coincides with the zero streamline which is determined from the integration of the radial

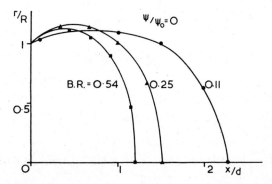

Fig. 3.13(a). Change of recirculation zone boundary with blockage ratio (for discs).

distribution of axial velocity components. The reverse flow zone is bounded by the zero velocity line and the boundaries of the reverse flow zone and recirculation zone coincide at the separation point on the bluff body and the downstream stagnation point. The centre or 'eye' of the eddy lies on the reverse flow boundary and is the point where the static pressure is a minimum.

The radial position of the velocity maximum converges towards the axis and is shown as the line where $d\bar{u}/dr = 0$. The forebody divergent angle α causes the flow to be deflected away from the bluff body and has the value zero for a cylinder with axis parallel to main flow and 180° for a disc. Davies and Beér[21] have made detailed measurements of mean velocities, static pressures and turbulent stresses for a range of forebody geometries and blockage ratios $(d/D)^2$. The width of the recirculation eddy (Fig. 3.13a) is hardly affected by change in blockage ratio. The main influence of variation of blockage ratio is to cause a reduction in its length from the maximum value of $L/d = 2.6$ measured by Carmody[22] for a disc in a free stream. Maximum deflection of the main stream is caused

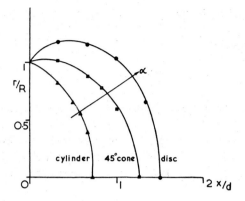

Fig. 3.13(*b*). *Change of recirculation zone boundary with forebody geometry* (*b.r.* = 0·25)

by a disc. Reduction of the forebody angle to a 45° cone and then to a cylinder results in a progressive reduction in the size of the recirculation zone (Fig. 3.13b).

The proportion of the flow from the main stream that is recirculated within the recirculation zone can be considered as a measure of the strength of the vortex and is of prime importance in determining the effectiveness for flame stabilisation. Reverse mass flow rates are determined at each axial station by

$$\dot{M}_r = \int_0^{r_0} 2\pi r \rho \bar{u} \, dr \qquad (3.34)$$

where r_0 is the radial position of the zero velocity boundary ($\bar{u} = 0$). Relative reverse mass flow rates are increased by increasing the blockage

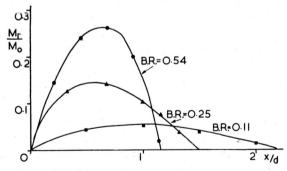

Fig. 3.14(*a*). *Effect of blockage ratio of axial distribution of reverse mass flow rate* (*for discs*).

ratio and reduced by reducing the angle of the forebody (Figs. 3.14a and
3.14b).

The spatial distribution of the mean stream function is calculated from

$$\psi = \int_0^\infty \bar{u} r \, dr \tag{3.35}$$

Fig. 3.15a shows the streamline pattern in the wake of a disc for a blockage
ratio of 0·25.

Turbulence characteristics have been measured[21] by using a constant
temperature hot wire anemometer. Because of the high levels of turbulence
and the large variation in flow directions, measurements were made at

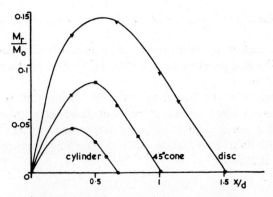

Fig. 3.14(*b*).   *Effect of forebody geometry on reverse mass flow rate* (*b.r.* = 0·25).

each point within the flow field with the hot wire placed at six different
angles to the flow. Analyses of these six measurements allowed direct
computation of the local turbulence intensity, the local kinetic energy of
turbulence and the turbulent shear stresses. Distributions of local turbulence intensity, $(\overline{v'^2})^{1/2}/\bar{u}$, local turbulence kinetic energy,

$$k = \frac{(\overline{u'^2} + \overline{v'^2} + \overline{w'^2})}{\bar{u}^2}$$

and the normalised shear stress, $\overline{u'v'}/\bar{u}_0{}^2$, for a disc with blockage ratio of
0·25 are shown in Figs. 3.15b, c and d respectively. The measured distri-
butions of the turbulence characteristics show that a region of high
turbulence intensity and shear exists within the recirculation zone. Regions
of highest turbulence intensity coincide with the 'eye' of the vortex where

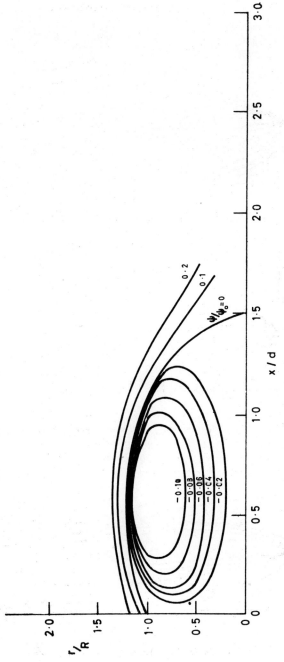

Fig. 3.15(a). Spatial distribution of normalised stream function (ψ/ψ_0) in the wake of a disc.

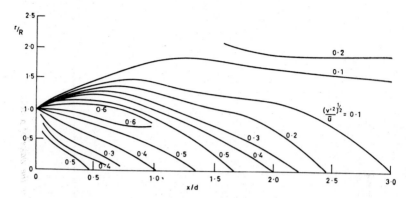

Fig. 3.15(*b*).   *Spatial distribution of* $\overline{(v'^2)}^{1/2}/\bar{u}$ *in the wake of a disc.*

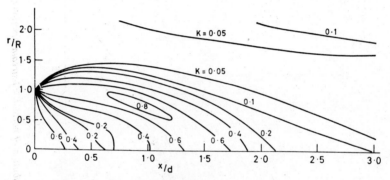

Fig. 3.15(*c*).   *Spatial distribution of local kinetic energy of turbulence (k) in the wake of a disc.*

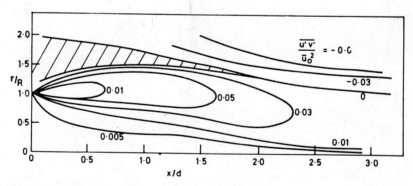

Fig. 3.15(*d*).   *Spatial distribution of normalised shear stress* $\overline{u'v'}/\bar{u}_0^2$ *in the wake of a disc.*

static pressures are minimum. The turbulence decays rapidly to free jet proportions within three disc diameters. All the results discussed above were made under isothermal room temperature conditions.

3.7 COMBUSTION IN RECIRCULATION ZONES

Combustion in the wake of stabilisers causes changes in the shape and strength of recirculation zones. Reaction within the recirculation zone, with the associated temperature increase, would result in an increased

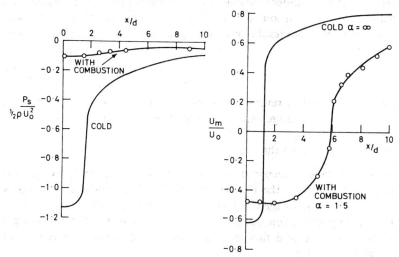

Fig. 3.16. *Effect of combustion on static pressure and axial velocity along the axis of the wake behind a 60° V-shaped flame holder (Bespalov[24]).*

volumetric flow rate of the order of five times if the shape and strength of the zone remained unaltered. Burning of fuel within the zone causes an increase in the length of the zone and an increase in the static pressure. Temperature and concentration distributions within the zone are very uniform and aerodynamic changes within the zone are directly dependent upon the temperature of the well mixed gases in the zone. Very few reliable measurements have been made within the zone under combustion conditions with the exception of those made by Winterfeld[23] and those reported by Bespalov.[24]

The measurements taken by Bespalov[24] in the wake of a 60° V-shaped flame holder show (Fig. 3.16) that static pressure increases from $-1 \cdot 1(\frac{1}{2}\rho U_0{}^2)$

without combustion to $-0\cdot1(\frac{1}{2}\rho U_0{}^2)$ with combustion. The length of the recirculation zone increases from $1\cdot6x/d$ to $6x/d$. The maximum reverse flow velocity remains unchanged, but the volumetric flow rate is increased by a factor of six due to combustion. With temperatures in the wake of the order of 1400°C the mass reverse flow rates remain the same under combustion conditions as under cold conditions.

The increase in static pressure and the increase in length are a direct result of thermal expansion of the gases and the pressure and dimensions of the zone depend upon the amount of heat released. Experiments show that as mixtures become leaner, static pressure and length of zone decrease towards the values under cold conditions. Some of the earliest comparisons of flow under cold and combustion conditions were made by Bovina[25] who measured the average residence time of particles in the reverse flow zone:

$$t = \frac{V}{S\bar{u}_{\mathrm{esc}}} \tag{3.36}$$

where V = volume of reverse flow zone, S = surface area of reverse flow zone, and $\bar{u}_{\mathrm{esc}}$ = average escape velocity of gas through the surface of the reverse flow zone.

Bovina[25] concluded that

(1) the average residence time of particles in the reverse flow zone is proportional to the dimensions of the flame holder and inversely proportional to the average approach flow velocity;

(2) during combustion, residence time is approximately 2–8 times larger than during cold flow and is virtually independent of change of mixture ratio;

(3) residence time decreases with increase of intensity of turbulence.

Residence time may be calculated from the following simple empirical formula:

$$t = k\frac{d}{U_0} \tag{3.37}$$

where d = diameter or characteristic height of flame holder, U_0 = approach flow velocity, and k = proportionality factor depending on the geometric shape of the flame holder.

For 30° V-shaped flame holders the value of k was found to be 37 during cold flow and 104 during combustion.

From knowledge of the average residence time and of the volume of the reverse flow zone, the rate of mass transfer between the reverse flow zone

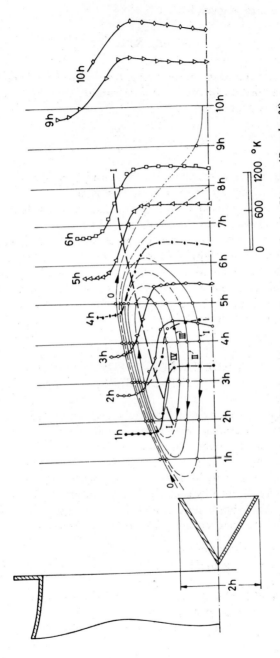

Fig. 3.17.   Temperature distribution and streamlines in the wake of a V-gutter (Bespalov [24]).

and the main flow is given by

$$\dot{m} = \bar{u}_{esc} S \bar{\rho} = \frac{V}{\bar{t}} \bar{\rho} \qquad (3.38)$$

The heat transferred from the reverse flow zone can also be determined from the composition of the mixture, assuming complete combustion in the reverse flow zone, from

$$\dot{Q} = \dot{m} \bar{c}_p \bar{T}_r = \frac{V}{\bar{t}} \bar{\rho} \bar{c}_p \bar{T}_r \qquad (3.39)$$

where $\bar{\rho}$, $\bar{c}_p$, $\bar{T}_r$ are average values of density, specific heat and temperature respectively within the reverse flow zone.

A number of experiments have shown that over a wide range of flow velocity and mixture ratio the temperature and chemical composition of gases in the reverse flow zone remain practically constant. The measurements of Bespalov are shown in Fig. 3.17. As critical blow off conditions are approached by increasing flow velocity, the size of the constant temperature zone is reduced. These experiments demonstrate that flame stabilisation is achieved by a continuous supply of heat carried by hot combustion products from the reverse flow zone into the main flow.

3.8 STABILITY LIMITS FOR BLUFF BODY STABILISERS

When the main stream velocity around a bluff body stabiliser is progressively increased, a critical value is reached at which the flame blows off. The limits of stable combustion are usually characterised by blow off and flashback velocities at a given mixture ratio. Maximum blow off velocities generally occur with stoichiometric mixture ratios and blow off velocities decrease as mixtures are made richer or leaner. Numerous experiments have been summarised by Bespalov[24] who shows that stability limits are widened under the following conditions:

(1) increase in dimensions of the stabiliser;
(2) increase in drag coefficient of the stabiliser (geometric form of the stabiliser);
(3) increase of final and initial temperature of the mixture;
(4) increase of pressure in the flow;
(5) decrease of intensity of turbulence in the main flow.

Stable combustion with liquid droplet mixtures is achieved with very lean mixtures well beyond the ignition limits of homogeneous mixtures of hydrocarbon vapours with air. This is explained by the fact that in a moving two-phase mixture fuel droplets settle on the surface of the stabiliser. Depending upon the temperature of the flame holder, the settled fuel is partially evaporated and partially atomised in the recirculation zone. As a result the fuel mixture is richer in the recirculation zone than in the main flow. With increase of flow velocity the fraction of drops settling on the flame holder increases, thereby allowing stabilisation at high flow velocities.

REFERENCES

1. Hirschfelder, J. O. and Curtiss, C. F., 'Theory of propagation of flames. Part I—general equations', Third Symposium on Combustion, Flames and Explosion Phenomena, p. 121, Williams and Wilkins, Baltimore, 1949.
2. Lewis, B. and von Elbe, G., *Combustion, Flames and Explosions of Gases*, Academic Press, New York, 1951; revised 1961.
3. Zeldovich, Y. B. and Frank-Kamenetsky, D. A., *Acta Physiochim, URSS* 1938, **9**, p. 341. *Compt. Rend. Acad. Sci., USSR* 1938, **19**, p. 693.
4. Friedman, R. and Burke, E., *J. Chem. Phys.* 1953, **21**, p. 710.
5. Egerton, A. C. and Lefebvre, A. H., 'Flame propagation: the effect of pressure variation on burning velocities', *Proc. Roy. Soc.* 1954, **A222**, p. 206.
6. Egerton, A. C., Saunders, O. A. and Spalding, D. B., 'The chemistry and physics of combustion', Inst. of Mechanical Engineers, Joint Conference on Combustion, pp. 14–33, 1955.
7. Wohl, K., Kapp, N. M. and Gazley, C., 'Stability of open flames', Third Symposium on Combustion, p. 3, Williams and Wilkins, Baltimore, 1949.
8. Williams, G. C., Hottel, H. C. and Scurlock, A. C., 'Flame stabilisation and propagation in high velocity gas streams', Third Symposium on Combustion, p. 21, Williams and Wilkins, Baltimore, 1949.
9. Damköhler, G., *Jahrb. deut. Luftfartforsch.* 1939, p. 113; *Z. Elektrochem.* 1940, **46**, p. 601. English translation, NACA Tech. Memo 1112, 1947.
10. Williams, D. T. and Bollinger, L. M., 'Effect of turbulence on flame speeds of bunsen-type flames', Third Symposium on Combustion, p. 176, Williams and Wilkins, Baltimore, 1949.
11. Shchelkin, K. I., *Zh. Tekhn. Fiz.* 1943, **13**, pp. 520–30. English translation, NACA Tech. Memo 1110, 1947.
12. Scurlock, A. C., 'Flame Stabilisation and Propagation in High Velocity Gas Streams', Meteor Report No. 19, Fuels Research Laboratory, MIT Press, Cambridge, Mass., 1948.
13. Lefebvre, A. H. and Reid, R., 'The influence of turbulence on the structure and propagation of enclosed flames', *Combustion and Flame* 1966, **10**, pp. 355–66.

14. Karlovitz, B., Denniston, D. W. and Wells, F. E., *J. Chem. Phys.* 1951, **19**, p. 541.
15. Scurlock, A. C. and Grover, J. H., 'Propagation of turbulent flames', Fourth Symposium on Combustion, pp. 645–58, Williams and Wilkins, Baltimore, 1953.
16. Summerfield, M., Reiter, S. H., Kebely, V. and Mascalo, R. M., *Jet Propulsion* 1954, **24**, p. 254.
17. Spalding, D. B., 'Theory of rate of spread of confined turbulent premixed flames', Seventh Symposium on Combustion, p. 595, Butterworths, London, 1959.
18. Wohl, K., Shore. L., von Rosenberg, H. and Weil, C. W., 'Burning velocity of turbulent flames', Fourth Symposium on Combustion, p. 620, Williams and Wilkins, Baltimore, 1953.
19. Wright, F. H. and Zukoski, E. E., 'Flame spreading from bluff-body flameholders', Eighth Symposium on Combustion, p. 933, Williams and Wilkins, Baltimore, 1962.
20. Zukoski, E. E. and Marble, F. E., *Combustion Researches and Reviews. AGARD*, p. 167, Butterworths, London, 1955.
21. Davies, T. W. and Beér, J. M., 'The turbulence characteristics of annular wake flow', Heat and Mass Transfer in Flows with Separated Regions, International Seminar, Hercig-Novi, Yugoslavia, 1969.
22. Carmody, J., 'Establishment of the wake behind a disc', *Trans. ASME*, **86D**, *J. Basic Eng.* 1964, **4**, p. 869.
23. Winterfeld, G., 'On processes of turbulent exchange behind flameholders', Tenth Symposium on Combustion, pp. 1265–75, The Combustion Institute, 1965.
24. Bespalov, I. V., 'Physical principles of the working process in combustion chambers of jet engines by Raushenbakh *et al.*, translated from Russian, Clearinghouse for Federal Scientific and Technical Information, USA, AD 658 372 p. 366, May 1967.
25. Bovina, T. A., 'Studies of exchange between the recirculation zone behind the flameholder and the outer flow', Seventh Symposium on Combustion, p. 692, Butterworths, London, 1958.
26. Karlovitz, B., Denniston, D. W., Knapschaffer, D. H. and Wells, F. E., Fourth Symposium on Combustion, p. 613, Williams and Wilkins, Baltimore, 1953.
27. Reed, S. B., *Combustion and Flame* 1967, **11**, p. 177.
28. Edmondson, H. and Heap, M. P., *Combustion and Flame* 1970, **14**, p. 191.

CHAPTER 4

Analysis—Prediction Methods

NOMENCLATURE

a	distance to apparent origin of jet.
c	specific heat.
d	nozzle diameter.
$\mathbf{d}$	mean flow rate of deformation tensor.
$\mathbf{F}$	body force per unit volume.
H	heat of combustion.
h	time–mean enthalpy.
$\mathbf{J}$	turbulent flux vector.
l	mixing length.
M	mean molecular weight.
m	time–mean chemical species' mass fraction.
P	point in flow field.
p	time–mean pressure.
R	mass rate of creation per unit volume (with subscript); universal gas constant (without subscript).
r	radial co-ordinate.
S	swirl number = angular momentum flux/axial momentum flux × nozzle radius.
T	temperature.
t	time.
$\mathbf{v}$	time–mean velocity.
z	axial co-ordinate.

Greek Symbols

Γ	turbulent exchange coefficient.
$\delta r, \delta z$	small distances in r- and z-directions.
θ	polar co-ordinate.
λ	mixing length parameter.
μ	turbulent viscosity.

85

ξ non-dimensional radial co-ordinate $= r/(z + a)$.
ρ time–mean density.
σ Prandtl–Schmidt number.
τ turbulent stress (momentum flux) tensor.
∇ vector differential operator.

Subscripts
fu relating to fuel.
h relating to enthalpy.
j relating to chemical species j.
m maximum value at a particular axial station.
max   position where $u/u_m = 0\cdot01$.
o   value at orifice of jet.
p relating to constant pressure.
rz, etc. rz-component of second-order tensor, etc.
z, r, θ components of vector in co-ordinate directions.
μ relating to turbulent viscosity.

Superscripts
        turbulent fluctuating component.

4.1   INTRODUCTION

One of the more general systems that can be considered theoretically is
the case of turbulent swirling flows with combustion. For flows in which
no recirculation takes place, boundary layer approximations may apply
which leads to considerable simplification of the analysis.

Flow systems are treated as being in a quasi-steady state, account
being taken of time–mean average properties of velocities, pressure,
temperature, chemical species concentrations and turbulence properties.
In turbulent diffusion flames, interaction of mixing and chemical reaction
add greatly to the complexity of the problem.

Prediction of time–mean average velocity, pressure, temperature and
concentration in turbulent combustion systems can be made provided the
turbulent momentum flux tensor τ, the turbulent enthalpy flux vector $\mathbf{J}_h$
and the turbulent chemical species' flux vector $\mathbf{J}_j$ (one for each chemical

species) are specified. Turbulent exchange coefficients, μ, Γ_h, and Γ_j (defined by analogy with Newton's, Fourier's and Fick's laws for laminar flows) are generally used, the latter two bearing fixed ratios (the Prandtl and Schmidt numbers) to the former.

The main difficulty in any prediction method is lack of knowledge of these turbulent fluxes and exchange coefficients. Turbulence models have been used to specify the fluxes directly or indirectly by way of the exchange coefficients, but their validity is questionable in swirling flows. Progress in this field has been hampered by the comparative scarcity of results reported in the literature of spatial distributions of velocity, temperature and chemical composition in swirling jets and in flames.

Until recently, another difficulty has been the computational one of solving the governing Reynolds equations when a hypothesis for the turbulent flows has been incorporated. Powerful numerical techniques are evolving[2,3,4] for solution of such systems of equations, but the incorporation of all the salient features of a particular flow configuration is still relatively complex. For systems with swirl, problems with entrainment, pressure gradients and grid stability have been encountered, whilst the additional complications of combustion generate sudden changes in density, temperature and composition.

Advances in measuring techniques have led to measurement of some time–mean quantities in turbulent combustion systems. The aim of an analysis is to provide new or additional information on the fundamental mixing and chemical–kinetic processes and to provide hypotheses about these processes. On the other hand, solution of the Reynolds equations for time–mean values is restricted through lack of knowledge of turbulence hypotheses and experimental verification of turbulence models is still required for prediction of mean flow patterns. Comparison of time–mean predictions with measured values allows improvements in hypotheses and turbulence models to be made, provided the solution procedure is sufficiently accurate.

4.2 THE GOVERNING EQUATIONS

BASIC DIFFERENTIAL EQUATIONS

The basic equations describe the flow of turbulent, chemically reacting, multicomponent mixtures, in which heat and mass transfer are occurring. The vector–tensor turbulent flux (Reynolds) equations of conservation of mass, momentum, enthalpy and chemical species are taken as

$$\frac{D\rho}{Dt} + \rho(\nabla \cdot \mathbf{v}) = 0 \tag{4.1}$$

$$\rho \frac{D\mathbf{v}}{Dt} = \rho \mathbf{F} - \nabla p + \nabla \cdot \boldsymbol{\tau} \tag{4.2}$$

$$\rho \frac{Dh}{Dt} = -\nabla \cdot \mathbf{J}_h - p(\nabla \cdot \mathbf{v}) + \nabla \mathbf{v} : \boldsymbol{\tau} \tag{4.3}$$

$$\rho \frac{Dm_j}{Dt} = -\nabla \cdot \mathbf{J}_j + R_j \tag{4.4}$$

There is an equation of type (4.4) for each chemical species present. In obtaining these equations, external force fields have been neglected and the internal energy has been equated with enthalpy. Only turbulent contributions to the fluxes are considered, the molecular contributions being considered negligibly small in fully turbulent flows.

If p, $\mathbf{v}$, T and m_j are considered as unknowns the equation system is not closed. Further unknowns are ρ, h and R_j and the turbulent fluxes $\boldsymbol{\tau}$, $\mathbf{J}_h$ and $\mathbf{J}_j$.

THERMODYNAMIC RELATIONSHIPS

Thermodynamic relationships provide some of the necessary extra equations to close the system; these equations are taken as

$$p = \rho \frac{R}{M} T \tag{4.5}$$

$$h = c_p T + \sum_j (H_j m_j) \tag{4.6}$$

$$R_j = -F_j \exp(-E_j/RT) \tag{4.7}$$

There is an equation of type (4.7) for each chemical species present. Equation (4.5) is the equation of state, and density variations are determined from temperature variations under the assumption that chemical composition has negligible effect on density and that pressure changes are caused solely through eqn. (4.2). Equation (4.6) demands that an average specific heat c_p be used for the mixture and that its value be independent of temperature. The rates of production R_j of chemical species j are usually provided by expressions of the Arrhenius type [eqn. (4.7)] where

F_j, the frequency (pre-exponential) factor, and E_j, the activation energy, are used. One fewer chemical species[1] differential equation of type (4.4) and one fewer rate equation of type (4.7) need be solved since the relation

$$\sum m_j = 1 \tag{4.8}$$

is always valid.

FLUX LAWS

The turbulent fluxes are related to correlations of turbulent fluctuations by the relations (neglecting density fluctuation contributions)

$$\boldsymbol{\tau} = -\rho \overline{\mathbf{v'v'}} \tag{4.9}$$

$$\mathbf{J}_h = \rho \overline{\mathbf{v'}h'} \tag{4.10}$$

$$\mathbf{J}_j = \rho \overline{\mathbf{v'}m'_j} \tag{4.11}$$

These fluxes are not known and it is their specification which closes the equation system and allows predictions to be made. By analogy with laminar flows, extensions of Newton's constitutive stress–strain relation, of Fourier's law of heat conduction and of Fick's law of diffusion have been postulated and used with variable turbulent exchange coefficients μ, Γ_h and Γ_j defined by

$$\boldsymbol{\tau} = 2\mu \mathbf{d} \tag{4.12}$$

$$\mathbf{J}_h = -\Gamma_h \nabla h \tag{4.13}$$

$$\mathbf{J}_j = -\Gamma_j \nabla m_j \tag{4.14}$$

The contribution of the turbulence energy to the normal components of $\boldsymbol{\tau}$ has been neglected for simplicity, for, in the subsequent analysis, interest is restricted to shear stress components.

Without assuming isotropic turbulence, the μ, Γ_h and Γ_j associated with different components of eqns. (4.12) to (4.14) require individual specification. A calculation procedure which assumes isotropy and, as a further simplification, assumes Γ_h and Γ_j to bear constant ratios (the Prandtl and Schmidt numbers σ_h and σ_j) to the isotropic turbulent viscosity, via the relations

$$\sigma_h = \frac{\mu}{\Gamma_h} \qquad \sigma_j = \frac{\mu}{\Gamma_j} \tag{4.15}$$

possesses the desirable property of having only one outstanding unknown,

μ, in the governing equation system. A simple turbulence model for μ would then close the system and allow predictions to be made. Many simple flows are amenable to this simplified treatment. However, for combustion work or swirling flows, this analysis is not sufficient and more elaborate non-isotropic models must be retained.

Turbulence Models

It is the unknown fluxes which lead to difficulties in satisfactorily solving the Reynolds equations for time–mean values. Predictions can only be made if they are specified in terms of the mean quantities already in the system or in terms of further unknowns with correspondingly further equations.

A turbulence model is some hypothesis which specifies the unknown fluxes and so closes the system and allows for solution for time–mean $\mathbf{v}$, h, T, p, ρ and m_j. This can take the form of specifying the fluxes directly or indirectly by way of the exchange coefficients. Several models of turbulence have been summarised by Spalding[1] together with an assessment of their merits and demerits. Each has been suitable for a particular flow configuration, the more recent ones evolving in attempts towards universality.

In Prandtl's mixing length hypothesis he assumed that the isotropic viscosity μ could be specified in terms of local gradients of mean quantities via a mixing length l and constant mixing length parameter λ. The mixing length l is taken as the constant λ, times the width of the mixing region, where λ takes a value somewhere between 0·007 and 0·13, depending on the situation.

It is usual to take the viscosity to be proportional to the second invariant of the mean flow rate of deformation tensor, as in elasticity, and an isotropic model is

$$\mu = \rho l^2 (2\mathbf{d}:d)^{1/2} \tag{4.16}$$

4.3   EQUATIONS IN THE CYLINDRICAL POLAR CO-ORDINATE SYSTEM

Cylindrical Polar Form of Equations

Using tensor theory, the basic equations of section 4.2 may be written in any co-ordinate system. The basic equations are written in the cylindrical polar co-ordinate system (z, r, θ) and the motion is assumed to be quasi-steady $(\partial/\partial t = 0)$ and axisymmetric $(\partial/\partial\theta = 0)$. Neglecting kinetic heating they become

$$\frac{\partial}{\partial z}(\rho v_z) + \frac{1}{r}\frac{\partial}{\partial r}(r\rho v_r) = 0 \tag{4.17}$$

$$\rho\left(v_z\frac{\partial}{\partial z}v_z + v_r\frac{\partial}{\partial r}v_z\right) = \frac{\partial}{\partial z}(\tau_{zz}) + \frac{1}{r}\frac{\partial}{\partial r}(r\tau_{rz}) - \frac{\partial p}{\partial z} \tag{4.18}$$

$$\rho\left(v_z\frac{\partial}{\partial z}v_r + v_r\frac{\partial}{\partial r}v_r - \frac{v_\theta^2}{r}\right) = \frac{\partial}{\partial z}(\tau_{rz}) + \frac{1}{r}\frac{\partial}{\partial r}(r\tau_{rr}) - \frac{\tau_{\theta\theta}}{r} - \frac{\partial p}{\partial r} \tag{4.19}$$

$$\rho\left(v_z\frac{\partial}{\partial z}v_\theta + v_r\frac{\partial}{\partial r}v_\theta + \frac{v_r v_\theta}{r}\right) = \frac{\partial}{\partial z}(\tau_{\theta z}) + \frac{1}{r^2}\frac{\partial}{\partial r}(r^2\tau_{r\theta}) \tag{4.20}$$

$$\rho\left(v_z\frac{\partial h}{\partial z} + v_r\frac{\partial h}{\partial r}\right) = -\frac{\partial}{\partial z}((J_h)_z) - \frac{1}{r}\frac{\partial}{\partial r}(r(J_h)_r) + \Phi \tag{4.21}$$

$$\rho\left(v_z\frac{\partial m_j}{\partial z} + v_r\frac{\partial m_j}{\partial r}\right) = -\frac{\partial}{\partial z}((J_j)_z) - \frac{1}{r}\frac{\partial}{\partial r}(r(J_j)_r) + R_j \tag{4.22}$$

where the dissipation function Φ is given by

$$\Phi = \nabla v : \tau = \tau_{zz}\frac{\partial v_z}{\partial z} + \tau_{rr}\frac{\partial v_r}{\partial r} + \tau_{\theta\theta}\frac{v_r}{r} + \tau_{rz}\left(\frac{\partial v_z}{\partial r} + \frac{\partial v_r}{\partial z}\right)$$

$$+ \tau_{r\theta}\left(\frac{\partial v_\theta}{\partial r} - \frac{v_\theta}{r}\right) + \tau_{z\theta}\left(\frac{\partial v_\theta}{\partial z}\right) \tag{4.23}$$

To obtain these, use has been made of the relations expressing D/Dt. $\mathbf{v}$, $\nabla \cdot \mathbf{v}$, τ and $\nabla v : \tau$ in the cylindrical polar co-ordinate system.
The component forms of the flux laws in this system are

$$\tau_{zz} = -\rho\overline{v'_z v'_z} \qquad \tau_{rz} = \tau_{zr} = -\rho\overline{v'_r v'_z}$$

$$\tau_{rr} = -\rho\overline{v'_r v'_r} \qquad \tau_{r\theta} = \tau_{\theta r} = -\rho\overline{v'_r v'_\theta}$$

$$\tau_{\theta\theta} = -\rho\overline{v'_\theta v'_\theta} \qquad \tau_{z\theta} = \tau_{\theta z} = -\rho\overline{v'_z v'_\theta} \tag{4.24}$$

$$(J_h)_z = \rho\overline{v'_z h'} \qquad (J_h)_r = \rho\overline{v'_r h'} \qquad (J_h)_\theta = \rho\overline{v'_\theta h'} \tag{4.25}$$

and

$$(J_j)_z = \rho\overline{v'_z m'_j} \qquad (J_j)_r = \rho\overline{v'_r m'_j} \qquad (J_j)_\theta = \rho\overline{v'_\theta m'_j} \tag{4.26}$$

The constitutive equations for the isotropic case are similarly written as

$$\tau_{zz} = 2\mu \frac{\partial v_z}{\partial z} \qquad \tau_{rz} = \tau_{zr} = \mu \left(\frac{\partial v_z}{\partial r} + \frac{\partial v_r}{\partial z}\right)$$

$$\tau_{rr} = 2\mu \frac{\partial v_r}{\partial r} \qquad \tau_{r\theta} = \tau_{\theta r} = \mu r \frac{\partial}{\partial r}\left(\frac{v_\theta}{r}\right) \qquad (4.27)$$

$$\tau_{\theta\theta} = 2\mu \frac{v_r}{r} \qquad \tau_{z\theta} = \tau_{\theta z} = \mu \frac{\partial v_\theta}{\partial z}$$

$$(\mathbf{J}_h)_z = -\Gamma_h \frac{\partial h}{\partial z} \qquad (\mathbf{J}_h)_r = -\Gamma_h \frac{\partial h}{\partial r} \qquad (\mathbf{J}_h)_\theta = 0 \qquad (4.28)$$

$$(\mathbf{J}_j)_z = -\Gamma_j \frac{\partial m_j}{\partial z} \qquad (\mathbf{J}_j)_r = -\Gamma_j \frac{\partial m_j}{\partial r} \qquad (\mathbf{J}_j)_\theta = 0 \qquad (4.29)$$

If isotropy is not assumed, different values of μ, Γ_h and Γ_j may be appropriate to different equations in (4.27) to (4.29).

The expression for eqn. (4.16) in cylindrical co-ordinates is

$$\mu = \rho l^2 \sqrt{2} \left[\left(\frac{\partial v_z}{\partial z}\right)^2 + \left(\frac{\partial v_r}{\partial r}\right)^2 + \left(\frac{v_r}{r}\right)^2 + \frac{1}{2}\left(\frac{\partial v_z}{\partial r}\right)^2\right.$$

$$\left. + \frac{1}{2}\left(r \frac{\partial}{\partial r}\left(\frac{v_\theta}{r}\right)\right)^2 + \frac{1}{2}\left(\frac{\partial}{\partial z} v_\theta\right)^2\right]^{1/2} \qquad (4.30)$$

BOUNDARY LAYER EQUATIONS

The equations given so far govern the flow of multicomponent, chemically reacting systems and solution of them with given boundary conditions determines the velocity, pressure, temperature and chemical species composition throughout the flow field. Such solution is non-trivial because of the abundance of terms in the equations and the many flux components about which little is known in turbulent flows. Their non-linear elliptic character demands that a lengthy numerical relaxation procedure be used in their solution.

Some physical flows may be simulated by a simplified form of these equations without much loss in accuracy. Such flows are called boundary layer flows and the application of boundary layer approximations results in truncation of the elliptic equations to parabolic form. Fewer terms and unknown fluxes are left in the equations and a simpler, quicker, forward-marching solution procedure can be applied. To qualify for this simplification, a flow must have a single predominant direction of flow and the flux components must be significant only in directions at right angles to this predominant direction. In particular, streamlines are not closed (there are no recirculation regions) and pressure variations should not allow downstream changes to influence upstream events.

Free, non-swirling jet flows qualify for boundary layer treatment; so do flows with weak swirl if the axial pressure gradient is still small and is not sufficient to cause recirculation. A strongly swirling jet flow with recirculation cannot be satisfactorily treated using reduced boundary layer equations and the full elliptic equations must be solved.

Boundary layer approximations are applied to weakly swirling flows by considering the relative orders of magnitude of terms assuming that

$$v_z, \ v_\theta \simeq 0(1) \qquad \frac{\partial}{\partial r} \simeq 0\left(\frac{1}{\varepsilon}\right) \qquad \text{and} \qquad \frac{\partial}{\partial z} \simeq 0(1)$$

where $\varepsilon \ll 1$.

The continuity equation then gives $v_r \simeq 0(\varepsilon)$ and the approximate order of magnitude of all the terms in the equations can be calculated in terms of ε, $\mathbf{\tau}/\rho$, $\mathbf{J}_h/\rho$ and $\mathbf{J}_j/\rho$. Assuming that turbulence fluctuations $\mathbf{v}'$, h', m'_j are $0(\varepsilon)$ and deleting terms of lowest order in each equation, the system reduces to

$$\frac{\partial}{\partial z}(\rho v_z) + \frac{1}{r}\frac{\partial}{\partial r}(r\rho v_r) = 0 \qquad (4.31)$$

$$r\left[\rho\left(v_z\frac{\partial v_z}{\partial z} + v_r\frac{\partial v_z}{\partial r}\right) + \frac{\partial p}{\partial z}\right] = \frac{\partial}{\partial r}(r\tau_{rz}) \qquad (4.32)$$

$$\rho\frac{v_\theta^2}{r} = \frac{\partial p}{\partial r} \qquad (4.33)$$

$$r^2\rho\left(v_z\frac{\partial v_\theta}{\partial z} + v_r\frac{\partial v_\theta}{r} + \frac{v_r v_\theta}{r}\right) = \frac{\partial}{\partial r}(r^2\tau_{r\theta}) \qquad (4.34)$$

$$r\rho\left(v_z\frac{\partial h}{\partial z} + v_r\frac{\partial h}{\partial r}\right) = -\frac{\partial}{\partial r}(r(\mathbf{J}_h)_r) \qquad (4.35)$$

$$r\left[\rho\left(v_z\frac{\partial m_j}{\partial z} + v_r\frac{\partial m_j}{\partial r}\right) - R_j\right] = -\frac{\partial}{\partial r}(r(\mathbf{J}_j)_r) \qquad (4.36)$$

There are now only four significant flux components. Their expressions in terms of turbulent fluctuations are contained in eqns. (4.24) to (4.26) and their constitutive counterparts are given in eqns. (4.27) to (4.29) (with the term $\partial v_r/\partial z$ omitted since it is two orders of magnitude smaller than $\partial v_z/\partial r$).

The boundary layer equivalent of the isotropic extension of Prandtl's model [eqn. (4.30)] is

$$\mu = \rho l^2 \left[\left(\frac{\partial v_z}{\partial r} \right)^2 + \left(r \frac{\partial}{\partial r} \frac{v_\theta}{r} \right)^2 \right]^{1/2} \tag{4.37}$$

When making predictions it is usual to assume isotropy but the little evidence that is available indicates that the assumption of isotropy is not valid for swirling flows, so that μ, Γ_h and Γ_j are not necessarily equal. The non-isotropic boundary layer equivalents of the constitutive assumptions [eqns. (4.27) to (4.29)] are

$$\tau_{rz} = \mu_{rz} \frac{\partial v_z}{\partial r} \qquad \tau_{r\theta} = \mu_{r\theta} \, r \frac{\partial}{\partial r} \frac{v_\theta}{r} \tag{4.38}$$

$$(\mathbf{J}_h)_r = -(\Gamma_h)_r \frac{\partial h}{\partial r} \qquad (\mathbf{J}_j)_r = -(\Gamma_j)_r \frac{\partial m_j}{\partial r} \tag{4.39}$$

Analogously, a non-isotropic extension of eqn. (4.37) is provided by

$$\mu_{rz} = \rho l_{rz}^2 \left[\left(\frac{\partial v_z}{\partial r} \right)^2 + \left(r \frac{\partial}{\partial r} \frac{v_\theta}{r} \right)^2 \right]^{1/2} \tag{4.40}$$

$$\mu_{r\theta} = \rho l_{r\theta}^2 \left[\left(\frac{\partial v_z}{\partial r} \right)^2 + \left(r \frac{\partial}{\partial r} \frac{v_\theta}{r} \right)^2 \right]^{1/2} \tag{4.41}$$

$$\lambda_{rz} = \frac{l_{rz}}{r_{\max}} \qquad \lambda_{r\theta} = \frac{l_{r\theta}}{r_{\max}} \tag{4.42}$$

Other generalisations of Prandtl's hypothesis are possible, for example,

$$\mu_{rz} = \rho l_{rz}^2 \left| \frac{\partial v_z}{\partial r} \right| \qquad \mu_{r\theta} = \rho l_{r\theta}^2 \left| r \frac{\partial}{\partial r} \frac{v_\theta}{r} \right| \tag{4.43}$$

Prandtl, Schmidt and (by analogy) $r\theta$-viscosity numbers are taken as

$$\sigma_h = \frac{\mu_{rz}}{(\Gamma_h)_r} \qquad \sigma_j = \frac{\mu_{rz}}{(\Gamma_j)_r} \qquad \sigma_{r\theta} = \frac{\mu_{rz}}{\mu_{r\theta}} \tag{4.44}$$

and, again, these need not be constant.

4.4 MACROSCOPIC BALANCES

INTEGRAL PROPERTIES

Integration of the basic equations (4.17), (4.18) and (4.20) over a section of the jet leads to

$$\frac{d}{dz} \int_0^\infty \rho r v_z \, dr = -(\rho r v_r)_\infty \qquad (4.45)$$

$$\frac{d}{dz} \int_0^\infty (\rho v_z{}^2 - \tau_{zz} + (p - p_\infty)) r \, dr = -(\rho r v_r v_z)_\infty \qquad (4.46)$$

$$\frac{d}{dz} \int_0^\infty (\rho v_z v_\theta - \tau_{\theta z}) r^2 \, dr = -(\rho r^2 v_r v_\theta)_\infty \qquad (4.47)$$

For a free jet system the following boundary conditions apply:

$$r = 0, \quad v_z \text{ finite}; \quad \frac{\partial v_z}{\partial r}, \, v_r, \, v_\theta = 0$$

$$p \text{ finite}; \quad \frac{\partial p}{\partial r} = 0, \, \boldsymbol{\tau} \text{ bounded} \qquad (4.48)$$

$$r \to \infty, \quad v_z, \, v_r, \, v_\theta, \, p, \, r^2, \, \tau_{r\theta}, \, \tau_{rz} \to 0$$

Equation (4.45) shows that the rate of increase of mass flux with axial distance is equal to the rate of mass entrainment or radial in-flow from large distances. The term on the right of eqn. (4.46) is zero, since $r v_r$ is finite and v_z is zero at infinity.

Hence,

$$\int_0^\infty (\rho v_z{}^2 - \tau_{zz} + (p - p_\infty)) r \, dr = G_z \qquad (4.49)$$

where G_z is the constant axial flux of axial momentum, comprised of momentum and pressure terms.

The term on the right of eqn. (4.47) is proportional to the product of the finite entrainment flux $- (2\pi \rho r v_r)_\infty$ per unit length of jet and the circulation $(2\pi r v_\theta)_\infty$ measured around the jet. For a swirling jet emerging into otherwise undisturbed surroundings, a further integration yields

$$\int_0^\infty (\rho v_z v_\theta - \tau_{\theta z}) r^2 \, dr = G_\varphi \qquad (4.50)$$

where G_φ is the constant axial flux of angular momentum.

For a boundary layer flow, it is usual to disregard the terms τ_{zz} and $\tau_{\theta z}$ in these expressions for G_z and G_φ.

Since G_z and G_φ are invariants in a swirling jet with zero circulation, they can be used to characterise the jet by defining a swirl number

$$S = \frac{G_\varphi}{G_z r_0}$$

where r_0 is the radius of the orifice.

The swirl number S does not completely characterise the jet, for the subsequent flow is also dependent on other factors such as nozzle geometry and confining walls. For example, a divergent duct encourages spreading, higher radial pressure gradients and can cause reverse flow.

ANALYSIS OF FLOW REGIMES

The non-swirling case, $S = 0$, has been adequately dealt with in much of the literature on fluid dynamics. Analyses have legitimately used boundary layer assumptions and neglected pressure variations.

As S increases, the cases of weak, moderate and strong swirling jets present themselves. For weak swirl (approximately $0 \le S \le 0.2$) the boundary layer equations can still be assumed to hold, the pressure variations being small and the coupling in the equations weak. For moderate swirl (approximately $0.2 \le S \le 0.5$) these equations including pressure variations are somewhat questionable but are still used, for present-day knowledge of turbulence necessitates similar approximation elsewhere in any analysis. A strongly swirling jet ($0.6 \le S$) possesses strong radial and axial pressure gradients in the region near the orifice. Evidently the reduced boundary layer equations of section 4.3 will not be valid and the above simplifying assumptions will not be applicable.

4.5 CALCULATION METHOD FOR DETERMINATION OF EXCHANGE COEFFICIENTS FROM MEAN VALUE DISTRIBUTIONS

As an intermediate contribution to the full prediction of turbulent swirling flows, a general inverse solution procedure has been developed by Lilley[5] for the case of non-recirculating swirling jets and flames. The method allows certain components of the turbulent fluxes and associated exchange coefficients to be calculated directly from limited experimental time–mean data, without the need for a complete solution to the problem. It is an intermediate step which is proving very useful, having been used on both swirling isothermal and combustion systems.[6,7] The method allows distributions of τ_{rz}, $\tau_{r\theta}(\mathbf{J}_h)_r$ and $(\mathbf{J}_j)_r$ and associated exchange coefficients (isotropy is not assumed) to be determined from the experimental mean distributions of v_z, v_θ, T and m_j. It thus provides a link between mean measurements and certain correlations of turbulent fluctuation components and throws light on the appropriateness or otherwise of any given turbulence model for the flow under consideration.

From the calculated values of turbulence quantities in swirling systems, modifications to turbulence models may be deduced. The resulting hypothesis can be incorporated into a general prediction procedure, so that the physical problem can be solved mathematically. The non-linear character of the Reynolds equations demands that a numerical technique be used on a high-speed computer, which is accurate, stable and economic.

CALCULATION PROCEDURE

Finite difference simulations of the differential equations (4.31) to (4.36) are required. Standard numerical formulae are used for differentiation and integration of any given function $y = y(x)$. For differentiation a three, five or seven point central difference formula is used over a series of points with constant x-interval. Integrations are performed using either the trapezoidal or Simpson's rule.

Any other integration or differentiation formulae may be used and there is no difficulty with regard to stability, convergency or accuracy of the formulae since they are used on smooth, experimentally fitted curves which are 'well behaved' in the sense that higher-order differences tend to zero as the order of the difference increases.

Assume now that curves have been fitted to experimentally observed time–mean axial and swirl velocities v_z and v_θ, temperature T and chemical species' concentrations m_j. Thus, for a given degree of swirl S, axial distance z and radial distance r, their values are easily calculated. The only unknowns in the reduced equation system are ρ, p, v_r, τ_{rz}, $\tau_{r\theta}$, $(\mathbf{J}_h)_r$ and $(\mathbf{J}_j)_r$. These can now be calculated at all points of the flow field.

Let P be a typical point with co-ordinates (z, r) and let $P(i,j)(1 \leq i, j \leq 7)$ be mesh points of a small 7×7 rectangular grid surrounding $P = P(4, 4)$, obtained by taking small increments δz and δr in the z- and r-directions. Thus the co-ordinates of $P(i, j)$ are

$$(z + (i - 4)\,\delta z, r + (j - 4)\,\delta r)$$

Since the values of v_z, v_θ, T and m_j [and hence h from eqn. (4.6)] are easily obtained at any node $P(i,j)(1 \leq i, j \leq 7)$, axial and radial derivatives of these are immediately calculable at the nodes $P(4, j)(1 \leq j \leq 7)$ and $P(i, 4)(1 \leq i \leq 7)$ respectively. The procedure for calculating the unknowns successively at P is as follows:

  (i) the density ρ is calculated from eqn. (4.5).
  (ii) equation (4.33) is integrated radially inwards to calculate p at the points $P(i, 4)(1 \leq i \leq 7)$. Hence $(\partial p / \partial z)$ is obtained at P.
  (iii) equation (4.31) is integrated radially outwards to calculate v_r at P.
  (iv) the values of all terms on the left-hand sides of eqns. (4.32), (4.34), (4.35) and (4.36) are calculated at P and values appropriate to the right-hand sides are deduced.
  (v) Integration of these values radially outwards gives τ_{rz}, $\tau_{r\theta}$, $(\mathbf{J}_h)_r$ and $(\mathbf{J}_j)_r$ at all radial points P at a given axial station z.

Iteration between steps (i) and (ii) to establish ρ and p converges rapidly, since the variation in p is small. The main use of calculating p is to establish $(\partial p / \partial z)$ for eqn. (4.32) which may not be small compared with the other terms in that equation. For all radial integration sweeps, N points P are used across the mixing layer, where N is quite large for increased accuracy. All derivatives are calculated by use of a three, five or seven point central difference formula and all integrations performed using Simpson's rule over three points. If experimental mean measurements of p and/or v_r are available, step (ii) and/or (iii) may be omitted. Boundary conditions are required for the integration stages; these are

$$p = p_\infty \qquad \text{at } r = \infty$$

$$v_r = \tau_{rz} = \tau_{r\theta} = (\mathbf{J}_h)_r = (\mathbf{J}_j)_r = 0 \qquad \text{at } r = 0$$

Repetition of the procedure at other axial stations z enables distributions of the rz- and $r\theta$-components of the Reynolds stress tensor $\mathbf{\tau}$, the r-component of the turbulent enthalpy flux vector $\mathbf{J}_h$ and the r-component of the turbulent chemical species' flux vector $\mathbf{J}_j$ to be obtained throughout the flow field. The relations between these and correlations of turbulent fluctuations are

$$\tau_{rz} = -\rho \overline{v'_r v'_z}, \qquad \tau_{r\theta} = -\rho \overline{v'_r v'_\theta} \qquad \qquad (4.51)$$

$$(\mathbf{J}_h)_r = \rho \overline{v'_r h'}, \qquad (\mathbf{J}_j)_r = \rho \overline{v'_r m'_j} \qquad \qquad (4.52)$$

The corresponding relations between the fluxes and the turbulent exchange coefficients are given in eqns. (4.38) and (4.39).

Any functions dependent on these can now be evaluated; for example.

mixing lengths l_{rz} and $l_{r\theta}$, mixing length parameters λ_{rz} and $\lambda_{r\theta}$ and Prandtl and Schmidt numbers σ_h and σ_j. An $r\theta$-viscosity number $\sigma_{r\theta}$ is also introduced [in eqn. (4.44)], by analogy with Prandtl–Schmidt numbers. For the non-isotropic generalisation of Prandtl's mixing length hypothesis used here, the lengths and parameters are determined from eqns. (4.40) to (4.42).

Other generalisations of Prandtl's hypothesis, with different mixing lengths, may be checked; for example, eqn. (4.43). The Prandtl, Schmidt and $r\theta$-viscosity numbers are calculated from eqn. (4.44).

Details of the computational procedure and the computer program are given in Ref. 5. Applying the calculation procedure to the measurements of mean distributions in flames with weak swirl, the distributions of turbulent exchange coefficients were calculated and shown to be functions of degree of swirl and position in the flow field. The turbulent viscosity was found to be highly anisotropic, the $r\theta$-component being an order of magnitude less than the rz-component. The variation of normalised μ_{rz} with swirl was found to be the opposite of that in the isothermal case—while μ_{rz} increases with swirl degree in the isothermal case, it decreases with swirl in the flame. This indicates that the effect of combustion was far more than just a density change.

REFERENCES

1. Spalding, D. B., 'Models of Turbulent Flow', Imperial College Department of Mechanical Engineering Report EF/TN/A/8, 1969.
2. Gosman, A. D., Pun, W. M., Runchal, A. K., Spalding, D. B. and Wolfshtein, M. W., *Heat and Mass Transfer in Recirculating Flows*, Academic Press, New York, 1969.
3. Patankar, S. V. and Spalding, D. B., *Intern. J. Heat Mass Transfer* 1967, **10**, p. 1389.
4. Kline, S. J., Morkovin, M. V., Sovran, G. and Cockrell, D. J. (eds.), Proceedings of Computation of Turbulent Boundary Layers—1968 AFOSR–IFP–Stanford Conference, 1968.
5. Lilley, D. G., Ph.D. Thesis, Department of Chemical Engineering and Fuel Technology, University of Sheffield, 1970.
6. Lilley, D. G. and Chigier, N. A., 'Nonisotropic turbulent stress distribution in swirling flows from mean value distributions', *Intern. J. Heat Mass Transfer* 1971, **14**, pp. 573–85.
7. Lilley, D. G. and Chigier, N. A., 'Nonisotropic exchange coefficients in turbulent swirling flames from mean value distributions', *Combustion and Flame* 1971, **16**, pp. 171–89.

CHAPTER 5

Swirling Flows

NOMENCLATURE

a	distance of the origin of the jet from the orifice.
B	axial width of channels.
C	empirical constant.
D	empirical constant.
d	orifice diameter.
E	empirical constant.
E_K	kinetic energy flux.
G	w_{m_o}/u_{m_o}.
G_x	axial flux of linear momentum.
$G_{x'}$	axial flux of linear momentum (excluding pressure term).
G_{φ}	axial flux of angular momentum.
g	acceleration due to gravity.
K	kinetic energy of turbulence.
k_p	static pressure error curve constant.
K_u	axial velocity error curve constant.
K_1	axial velocity decay constant.
K_2	swirl velocity decay constant.
K_3	pressure decay constant.
K_e	entrainment constant.
l	length scale of turbulence.
L	visible flame length.
m	mass flow rate.
m_o	mass flow rate at orifice.
$\dot{M}$	mass flow rate.
n	exponent in tangential velocity distribution equation.
p	static pressure.
p_{∞}	static pressure of the surroundings.
p_m	static pressure on the axis.
Q	volumetric flow rate.
Q_o	volumetric flow rate at the orifice.

R, R_1	radius of orifice.
Ri	Richardson number.
Ri*	modified Richardson number.
r	radial co-ordinate.
S	Swirl number $= G_\varphi/G_x R$.
t_s	residence time in stirred section.
$\bar{t}$	mean residence time.
T.I.	Turbulence intensity $\overline{(u'^2)}^{1/2}/U$.
u, U	axial component of the velocity vector.
u_m	axial velocity maximum.
u_{m_o}	axial velocity maximum at the orifice.
$\mathbf{W}$	velocity vector.
v	radial component of the velocity vector.
w, W	tangential component of the velocity vector.
w_m	tangential velocity maximum.
w_{m_o}	tangential velocity maximum at the orifice.
u', v', w'	turbulent fluctuations of the axial, radial and tangential components of velocity respectively.
x	axial co-ordinate.
z	number of vanes in cascade.

Greek Symbols

α	half jet angle ($u/u_m = 0.5$); vane angle.
Γ	circulation.
δ	coefficient depending upon type of swirl.
ε	ratio of radial to axial length scales for a non-swirling jet; efficiency of swirl generation.
ϑ	coefficient of kinetic energy flux.
μ_{eff}	effective viscosity.
ν	kinematic viscosity.
ξ	$r/(x + a)$; angle of adjustment of swirler.
ρ	density.
σ	$= \bar{W}_1/\bar{V}_1$.
τ	shear stress.
ψ	scalar potential function; blockage factor.
ω, Ω	angular velocity.

Subscripts

c	cold; centre
e	entrained; equivalent.

h   hub; hot.
m   maximum.
o   ambient; outer edge of swirler; initial; average.
tan   tangential.
tot   total.

5.1 INTRODUCTION

Swirling jets are used as a means of controlling flames in combustion chambers and have also found application in various types of spray driers and burners. The aerodynamics of the swirling turbulent jet combines the characteristics of rotating motion and the free turbulence phenomena encountered in jets and wake flows.

When rotating motion is imparted to a fluid upstream of an orifice, the fluid flow emerging from the orifice has a tangential velocity component in addition to the axial and radial components of velocity encountered in non-swirling jets. The presence of the swirl results in the setting up of radial and axial pressure gradients which, in turn, influence the flow field. In the case of strong swirl, the adverse axial pressure gradient is sufficiently large to result in reverse flow along the axis and the setting up of an internal recirculation zone.

While swirl burners are not new to combustion, it is only recently that a concentrated effort has been made to understand how and why rotating flow has such an important influence on the stability and combustion intensity of flames. Much information has come from a series of systematic trials in the Ijmuiden tunnel furnaces[1,2,6] and also from experimental and theoretical research carried out at universities and research establishments working in co-operation with the IFRF.[11,16,20,23,26,27,36] For purposes of this discussion, swirling turbulent systems are subdivided into the following groups: turbulent swirling jets with weak swirl; strongly swirling jets with internal recirculation; and buoyant turbulent jets in rotating flow fields. Within each of these groups cases are treated in which there is a density difference between jet flow and surroundings and also cases in which the jet flow is a turbulent diffusion flame.

When air is introduced tangentially into a burner, it is forced to change direction so that a spiral form of flow is set up. A balance is created between the centrifugal forces acting on the fluid particles and the pressure forces exerted on the walls of the tube. This force balance was demonstrated by measurements of the static pressure distributions in a burner,[1] where good agreement was obtained with the pressure distributions calculated from the measured tangential velocity distributions satisfying the equation $dp/dr = \rho W^2/r$. The low pressure in the central core of the rotating flow steadily recovers as the jet emerges from the burner. This results in an adverse axial pressure gradient so that, at sufficiently high

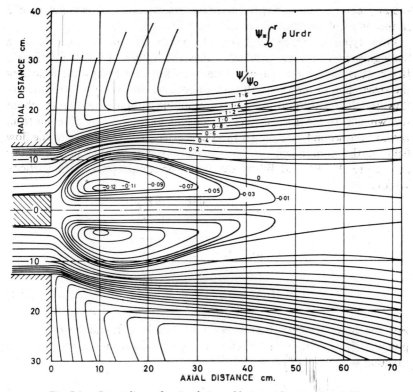

Fig. 5.1. Streamlines of recirculation eddy in swirling jet, $S = 1{\cdot}57$.

degrees of swirl, the flow reverses its direction and a central torroidal vortex is set up. The streamlines of the flow in this vortex region, shown in Fig. 5.1, were calculated from the measured velocity distributions. The presence of the torroidal vortex causes the outer boundaries of the jet to expand rapidly soon after the jet emerges from the burner exit. This initial expansion is not connected with entrainment of air from the surroundings. The length of the torroidal vortex core, as defined by the distance from the burner exit to the point of reversal of flow direction, increases with increase in the degree of swirl.[2] The particles of fluid emerging from the burner may be considered to be released from the constraining force exerted by the walls of the burner. They would thus have a tendency to fly out tangentially in a manner similar to that of an object rotated at the end of a string and then released. Particles of fluid are however restrained by the viscous forces and are also subjected to radial pressure forces. It is the combination of these effects which results in the shape of the streamlines shown in Fig. 5.1.

Flow types in which fluid particles move on circular paths can be distinguished according to whether they are 'rotational' or 'irrotational'. If a float is placed into water in rotational flow, the float while following the streamline in the flow will rotate about its own axis. In 'irrotational' flow, the float will follow a circular streamline without any rotation about its own axis. In the first case the vorticity vector has a value different from zero, in the second case the vorticity vector is zero everywhere in the flow field. If the components of the velocity vector $\mathbf{W}$ in the directions x, y, z are W_x, W_y and W_z we can write for components of the vorticity vector:

$$(\text{Curl } W)_x = \frac{\partial W_z}{\partial y} - \frac{\partial W_y}{\partial z}$$

$$(\text{Curl } W)_y = \frac{\partial W_x}{\partial z} - \frac{\partial W_z}{\partial x}$$

$$(\text{Curl } W)_z = \frac{\partial W_y}{\partial x} - \frac{\partial W_x}{\partial y} \tag{5.1}$$

For the homogeneous, frictionless fluid the irrotational flow is called potential flow because the velocity can be deduced from a scalar potential function $\psi = \psi(r)$ as

$$W_x = \frac{\partial \psi}{\partial x} \qquad W_y = \frac{\partial \psi}{\partial y} \qquad \text{and} \qquad W_z = \frac{\partial \psi}{\partial z} \tag{5.2}$$

The flow thus specified is irrotational because

$$(\text{Curl } W)_x = \frac{\partial W_z}{\partial y} - \frac{\partial W_y}{\partial z} = \frac{\partial^2 \psi}{\partial z \, \partial y} - \frac{\partial^2 \psi}{\partial y \, \partial z} = 0$$

According to potential theory, the necessary and satisfactory condition for flow to classify as potential flow is that the value of a line integral taken between two points in the flow field is independent of the path between the two points.

It follows that, along a closed line

$$\Gamma = \oint W\,ds = 0 \tag{5.3}$$

This line integral is called *circulation*.

When the centre of the system is enclosed by the closed line along which the line integral is taken, the circulation is non-zero but is of constant value. For example, along a circle of radius r

$$\Gamma = 2r\pi\,\frac{C}{r} = 2\pi C = \text{const} \tag{5.4}$$

This value of the circulation is the vorticity in the singular point, the centre of the potential vortex, and hence

$$W = \frac{C}{r} = \frac{\Gamma}{2\pi}\frac{1}{r} \tag{5.5}$$

As the streamline pattern can be derived from the potential corresponding to irrotational flow, this system is called the 'potential vortex'.

For $r = 0$, the velocity distribution $W = C/r$ would give $W = \infty$. The flow has a physical meaning only if it has a central core of finite radius. This core may exist either in the form of the fluid in solid body rotation or as a less dense fluid which does not take part in the rotation (*e.g.* vapour in the centre of rotating liquid).

Neglecting viscous forces, there is a balance between pressure forces and inertial forces in rotating flow. This can be expressed as

$$\frac{dp}{dr} = -\rho\,\frac{W^2}{r} \tag{5.6}$$

Substituting into the above equation for the velocity W from the relationships expressing the velocity distribution in free vortex and forced vortex (solid body rotation) flows we have, for the free vortex [eqn. (5.5)],

$$\frac{dp}{dr} = -\rho\left(\frac{\Gamma}{2\pi}\right)^2\frac{1}{r^3} \tag{5.7}$$

and

$$p - p_0 = \frac{\rho}{2}\left(\frac{\Gamma}{2\pi}\right)^2\frac{1}{r^2} \tag{5.8}$$

where p_0 is the ambient pressure.

For the forced vortex flow where

$$W = c'r \tag{5.9}$$

$$\frac{dp}{dr} = -\rho(c'r)^2 \frac{1}{r} = -\rho c'^2 r \tag{5.10}$$

and

$$p - p_c = -\tfrac{1}{2}\rho c'^2 r^2 \tag{5.11}$$

Thus the pressure increases in the core of a vortex proportionally with the square of the radius. Outside the core, in the free vortex flow region, the radial pressure distribution is hyperbolic.

The general characteristics of vortices are summarised in Table 5.1.

TABLE 5.1

	Forced vortex (solid body rotation)	Free vortex (potential vortex)	Combined vortex (Rankine vortex)
Tangential velocity distribution	$W = c'r$	$W = \dfrac{C}{r}$	$W = \dfrac{C'}{r}\left[1 - \exp\left(-\dfrac{r^2}{r_0^2}\right)\right]$
Angular velocity	$\Omega = $ const	—	—
Circulation	$2\pi\Omega r^2$	$2\pi C$	$2\pi C'\left[1 - \exp\left(-\dfrac{r^2}{r_0^2}\right)\right]$
Vorticity	$4\pi\Omega = $ const	0	$\dfrac{4\pi C'}{r_0^2}\left[\exp\left(-\dfrac{r^2}{r_0^2}\right)\right]$

5.2 THE SWIRL NUMBER, A NON-DIMENSIONAL CHARACTERISTIC OF ROTATING FLOW

In swirling free jets or flames, both axial flux of the angular momentum G_φ and the axial thrust G_x are conserved. These can be written as

$$G_\varphi = \int_0^R (Wr)\rho U 2\pi r\, dr = \text{const} \tag{5.12}$$

$$G_x = \int_0^R U\rho U 2\pi r\, dr + \int_0^R p 2\pi r\, dr = \text{const} \tag{5.13}$$

where U, W and p are the axial and tangential components of the velocity and static pressure respectively in any cross section of the jet. Since both these momentum fluxes can be considered to be characteristic of the aerodynamic behaviour of the jet, a non-dimensional criterion based on these quantities was recommended[2] as a criterion of swirl intensity as

$$S = \frac{G_\varphi}{G_x R} \qquad (R = \text{exit radius of the burner nozzle}) \qquad (5.14)$$

Subsequent experiments have shown that the swirl number S was the significant similarity criterion of swirling jets produced by geometrically similar swirl generators. Other similarity criteria, such as that by Thring and Newby, which take account of non-isothermal conditions and of confinement of jet flow by walls can also be applied in conjunction with the swirl number.

5.3 GENERATION AND CHARACTERISATION OF SWIRL†

METHODS OF SWIRL GENERATION

Methods of inducing rotation in a stream of fluid can be divided into three principal categories:

(a) tangential entry of the fluid stream, or of a part of it, into a cylindrical duct;
(b) the use of guide vanes in axial tube flow;
(c) rotation of mechanical devices which impart swirling motion to the fluid passing through them. This includes rotating vanes or grids and rotating tubes.

For the design of air registers, (a) and (b) are generally used in practice. Method (c) has sometimes been applied for experimental investigations of swirling jets.

Figure 5.2 represents a swirl generator with four tangential entries into a cylindrical duct. The advantages of this design are its simplicity and the fact that the intensity of swirl can be varied continuously between zero and a maximum value. Part of the combustion air, variable from zero to 100%, can be introduced through the tangential ports and the rest axially into the inner burner tube. The angular momentum flux of the swirling jet can then be controlled by varying the proportion of air introduced tangentially.

Another swirl generator developed at Ijmuiden is the 'movable-block swirl generator', shown in Fig. 5.3, which consists of two annular plates (P1 and P2) and two series of interlocking wedge-shaped blocks (B1 and

† The discussion in section 5.3 is based on a paper presented by Beér and Leuckel.[36]

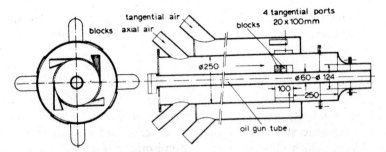

Fig. 5.2.   Scheme of swirl burner with axial and tangential air entries.

B2), each attached to one of the plates. Interlocked, the blocks form
alternate radial and tangential flow channels, such that the air flow
splits into an equal number of radial and tangential streams which combine further downstream into one swirling flow. By simply rotating the
back plate P2, the front plate P1 being fixed to the burner, radial channels
are progressively closed and tangential channels opened (or vice versa) so
that the resulting flux of angular momentum increases (or decreases
respectively) continuously, between zero and a maximum value, the latter
of which depends on the air flow rate and geometry of the swirl generator.
Advantages of this device are that it makes changes of swirl intensity
during burner operation possible, that the air pressures required for

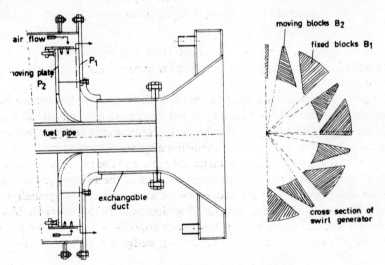

Fig. 5.3.   Scheme of moving-block swirl burner.

producing a certain swirl level are relatively low, and that it enables higher swirl numbers to be obtained.

A simple way of swirl generation as used for several industrial burners is by a single tangential entry into a cylindrical tube.[7-9] Its disadvantages are the bad symmetry of the flow distribution about the axis, and a low energetic efficiency of the swirl generation process. Nowadays, swirl registers of practical burners are mostly equipped with a cascade of guide vanes placed either into a radial flow [10-12] or in an axial tube flow.[13-15] The variation of swirl intensity in these burners requires a mechanical device for adjusting the vane direction.

CALCULATION OF THE SWIRL NUMBER S' FOR VARIOUS TYPES OF SWIRL GENERATOR

The calculation of the swirl number using eqns. (5.12)–(5.14) requires accurate measurements of velocity and of static pressure distributions to be made in a cross section of the swirling jet. The designer has sometimes no access to experimental data and there is interest in determining the swirl number directly from air register design data. Whereas G_φ and the velocity term in the expression of G_x can be predicted with reasonable accuracy from input data of various types of swirl generators, it is more difficult to predict the value of the static pressure integral [the second term in eqn. (5.13)] because the value of this latter term undergoes changes along the flow in the swirl generator depending upon its geometry (e.g. divergent nozzle extension). When the swirl number is calculated from input velocity distributions in the swirl generator rather than from the velocity distribution in the jet, the static pressure term can be omitted and the swirl number can be given with good approximation as

$$S' = \frac{G_\varphi}{G'_x R} \qquad (5.14a)$$

where

$$G'_x = 2\pi \int_0^R \rho U^2 r \, dr \qquad (5.13a)$$

and U and W represent axial and tangential components of the velocity in the swirl generator.

For guide-vane cascades in a radial flow (Fig. 5.4), the angular momentum can be expressed as

$$G_\varphi = \sigma \frac{\dot{M}^2}{\rho 2\pi B} \qquad (5.12a)$$

where $\dot{M}$ is mass flow rate, B is axial width of channels and σ is the ratio of the average tangential and radial velocity components at the swirler exit (radius R_1).

Now

$$\sigma = \frac{\bar{W}_1}{\bar{V}_1} \qquad (5.15)$$

where

$$\bar{W}_1 = \frac{G_\varphi}{\dot{M}R_1} \qquad (5.16)$$

and

$$\bar{V}_1 = \frac{\dot{M}}{\rho 2\pi R_1 B} \qquad (5.17)$$

Assuming the Reynolds number influence to be negligible, σ depends only upon the geometrical dimensions of the guide-vanes in the axis perpendicular cross section (see Fig. 5.4). It can then be shown from simple considerations[16,17] that

$$\sigma = \frac{1}{1 - \psi} \frac{\tan \alpha}{1 + \tan \alpha \tan (\pi/z)} \qquad (5.15a)$$

where $\psi = z\,s/2\pi R_1 \cos \alpha$ is a blockage factor which accounts for the finite thickness s of the vanes, and z is the number of vanes in the cascade.

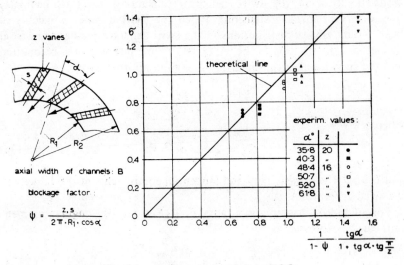

Fig. 5.4.   *Guide vane cascade in radial flow.*

The diagram on the right in Fig. 5.4 compares values of σ as determined by experiment[17] according to eqn. (5.12a) with those predicted from eqn. (5.15a). It can be seen that for larger angles α, values of σ are somewhat overestimated.

When applying a guide vane cascade in an axial tube flow with the vanes mounted in a hub of radius R_h and extended to the tube wall ($r = R$), as shown in Fig. 5.5, the angular momentum can generally be given as

$$G_\varphi = \int_{R_h}^{R} UW\rho 2\pi r^2 \, dr \qquad (5.12b)$$

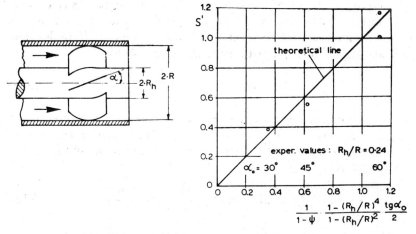

Fig. 5.5. Vane-type swirler in an axial tube flow.

For very thin vanes of constant chord and angle α, and provided that the axial velocity distribution is uniform over the tube cross section, this equation reduces to[18]

$$G_\varphi = 2\pi\rho U_0{}^2 \tan \alpha \, \frac{R^3 - R_h{}^{\sim}}{3} \qquad (5.12c)$$

Introducing the linear momentum

$$G'_x = \pi\rho U_0{}^2 (R^2 - R_h{}^2) \qquad (5.13b)$$

one may further write

$$G_\varphi = G'_x \tan \alpha \, R \, \frac{2}{3} \frac{1 - (R_h/R)^3}{1 - (R_h/R)^2} \qquad (5.12d)$$

Thus the swirl number of an annular swirler with constant vane angle α can be given as

$$S' = \frac{G_\varphi}{G'_x R} = \frac{2}{3}\left[\frac{1 - (R_h/R)^3}{1 - (R_h/R)^2}\right]\tan\alpha \qquad (5.14b)$$

and that of a hubless swirler as

$$S' = \frac{2}{3}\tan\alpha \qquad (5.14c)$$

Because of insufficient experimental data on vane-type swirl generators it is not possible to check the accuracy of swirl numbers predicted by eqns. (5.12a), (5.14b) and (5.14c). Some experimental results are available, however, for annular swirlers with long helicoidal vanes.[17,19] In this case, the vane angle α increases with the radius according to

$$\tan\alpha = \frac{r}{R}\tan\alpha_0 \qquad (5.18)$$

where α_0 represents the vane angle at the outer edge of the swirler. Again assuming uniform distribution of the axial velocity component over the tube cross section after the swirler, the theoretical equation of the angular momentum is[17]

$$G_\varphi = 2\pi\rho U_0{}^2 \tan\alpha_0 \frac{1}{1 - \psi}\frac{R^4 - R_h{}^4}{4R} \qquad (5.12e)$$

which leads to the following expression for the swirl number:

$$S' = \frac{G}{G'_x R} = \frac{1}{1 - \psi}\left(\frac{1}{2}\right)\frac{1 - (R_h/R)^4}{1 - (R_h/R)^2}\tan\alpha_0 \qquad (5.14d)$$

where ψ is again a blockage factor which allows for the finite thickness of the vanes.

Figure 5.5 compares experimental values of S', as determined for three different values of α_0, with the theoretical curve representing eqn. (5.14d). The agreement is reasonably good considering that the condition of uniform flow distribution after the swirler is only approximately satisfied.

TANGENTIAL ENTRY AND MOVABLE BLOCK-TYPE SWIRL GENERATORS

Detailed aerodynamic measurements were carried out at Ijmuiden[17] on both tangential entry and movable block-type swirl generators with variable swirl as described above.

The experimental data were treated to give:

(a) the total angular momentum flux as a function of the swirler adjustment;
(b) the type of swirl produced, *i.e.* the radial profiles of the axial and tangential components of velocity;
(c) the pressure required for the air supply and, as a part of it, the energy loss accompanying the process of swirl generation.

For the movable block swirler (Fig. 5.3) it was shown[17] that the angular momentum flux, expressed in terms of the coefficient σ [see eqn. (5.12a)] as a function of the swirler adjustment ξ/ξ_m, can be represented reasonably well by the following theoretical relationship:

$$\sigma = \frac{2\pi}{z\xi_m} \sin\alpha \frac{\cos\alpha \, [1 + \tan\alpha \tan(\xi/2)](\xi/\xi_m)}{\{1 - [1 - \cos\alpha \, (1 + \tan\alpha \tan(\xi/2))]\xi/\xi_m\}} \qquad (5.15b)$$

where ξ is the angle of adjustment of the swirler $(0 < \xi < \xi_m)$. The symbols of the geometrical dimensions used in this equation are explained in Fig. 5.6, which also shows a comparison of the theoretical curve with experimental results, obtained with three different mass flow rates. The maximum values of σ are substantially higher than those obtained with the vane-type swirler (Fig. 5.4).

For any given value of the coefficient σ, the swirl number S' of the flow

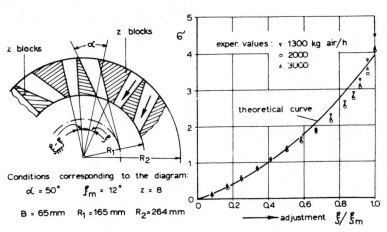

Fig. 5.6. *Movable block-type swirl generator.*

through a cylindrical or annular duct attached to a vane-type, or a movable block-type, swirler depends on the inner and outer radii (R_h and R) of the duct. Provided that the axial velocity distribution is uniform, and that the losses of angular momentum between swirler exit and duct are negligible, the relationship can be given as

$$S' = \frac{G_\varphi}{G'_x R} = \sigma \frac{R}{2B} \left[1 - \left(\frac{R_h}{R} \right)^2 \right] \tag{5.14e}$$

Hence the swirl number is directly proportional to the ratio R/B. Figure 5.7 represents an example of the variation of the swirl number as a function of the adjustment ξ/ξ_m of the movable block swirler, as measured for two different duct diameters. In the range covered by the experiments, the control curve proved to be independent of the mass flow rate, *i.e.* of the

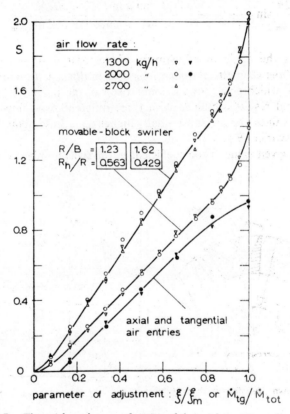

Fig. 5.7.   The swirl number as a function of the swirl generator adjustment.

Reynolds number. For comparison, the swirl number control curve for the burner with axial and tangential air inlets (Fig. 5.2) is also given in Fig. 5.7. In this case, the ratio between the tangential and the total mass flow rates ($\dot{M}_{tan}/\dot{M}_{tot}$) is the parameter of adjustment. It can be seen from Fig. 5.7 that, apart from its easy and quick adjustment of swirl intensity, the movable block swirler offers the advantage of substantially higher maximum swirl insities being obtainable.

THE EFFICIENCY OF SWIRL GENERATION

The pressure required for the combustion air to flow through a swirl register is a burner characteristic of great practical importance. Only part of the pressure drop across the burner reappears as kinetic energy of the swirling jet flow. The remainder of the pressure drop is due to mechanical energy losses which occur inside the air register. Considering first the kinetic energy flux through the burner throat, this can be expressed as

$$\dot{E}_k = \frac{\dot{M}}{2}\,\bar{u}^2(1 + \delta S^2) \tag{5.19}$$

where $\bar{u}$ is the mean axial velocity in the throat section and δ is a coefficient which depends on the type of swirl, the ratio of the inner and outer radii of the throat section, R_h/R, and the axial velocity distribution if not uniform. Assuming uniform axial velocity distribution, curves of $\vartheta = f(R_h/R, n)$ as

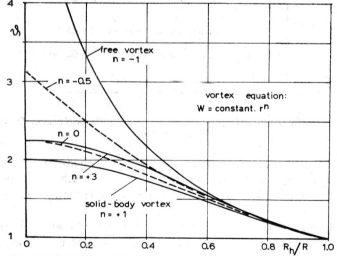

Fig. 5.8. *The coefficient of the kinetic energy flux of an annular swirling flow.*

plotted on Fig. 5.8 can be calculated from the relationship between swirl velocity and static pressure distributions for different swirl types. It can be seen that, for any given swirl number, the solid-body vortex ($n = 1$) represents the case of minimum kinetic energy and the free vortex ($n = -1$) gives the maximum kinetic energy. The constant tangential velocity

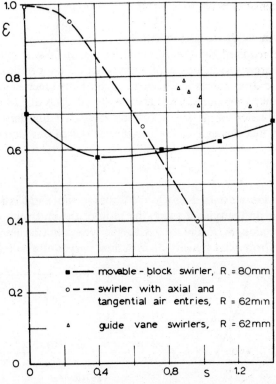

Fig. 5.9.   The efficiency of swirl generation as a function of the swirl number for different swirl generators.

vortex ($n = 0$) is the intermediate case between solid body and free vortex, and the case with the angular momentum flux very much concentrated in the outer part of the flow ($n = 3$) yields only marginally higher values of ϑ than those calculated for solid body-type rotation.

An expression of the 'efficiency of swirl generation' ε was defined[17] as the ratio of the flux of kinetic energy of the swirling flow through the burner throat to the drop of static pressure energy between the air inlets and the throat of the burner. Figure 5.9 shows the efficiency ε as a function of the swirl number S, as determined experimentally for different swirl generators. The curve of the movable block swirler has a minimum of 58% at intermediate swirl levels and an increase of ε towards higher swirl numbers which can be explained from theory. Values of ε obtained from various radial-flow guide vane swirlers at swirl numbers about 1 are between 70 and 80%. Compared to this, the air register with axial and tangential air entries was found to be inefficient for producing high swirl intensities: its efficiency was dropping rapidly with increasing swirl number and was only 40% at $S = 1$. This behaviour is due mainly to the large internal surface area of the inner burner pipe, in particular upstream from the tangential ports (Fig. 5.2).

Measurement data on the efficiency of swirl generation for vane swirlers in axial tube flow have been reported by Mathur and Maccallum.[19] It was shown that efficiencies of this type of swirl generator are generally low ($\varepsilon \simeq 0.3$ for $S \simeq 1$).

5.4   WEAK SWIRL ($S < 0.6$)

In swirling systems with weak swirl the axial pressure gradients are insufficiently large to cause internal recirculation. Swirl has the effect of increasing the rate of entrainment and the rate of velocity decay. Axial velocity profiles remain Gaussian until a swirl number of 0.5 and subsequently axial velocity maxima are displaced from the jet axis resulting in a double hump-type velocity profile.

For a tangential injection swirl generator, it has been shown[20] that similarity of the velocity profiles exists up to a swirl number of $S = 0.60$. The axial velocity was shown to fall into two patterns as seen from Fig. 5.10. For values of S up to 0.416 the form of the profiles is Gaussian while for S greater than 0.5 the velocity maximum is displaced from the jet axis and, at swirl numbers above 0.6, reverse flow commences. It can be seen from Fig. 5.10 that similarity of profiles is achieved as close as $x/d = 2$ to the nozzle for the lower swirl cases. This is closer to the nozzle than for the case of the non-swirling jet due to faster breakdown of the potential core. For the higher swirl numbers where the velocity maximum is displaced from the axis, similarity was not found since, in the initial region, the central velocity increases with x while U_m decreases. Similarity is ultimately achieved after the velocity maxima have converged on to the jet axis. The distance from the nozzle exit to the apparent origin, determined

S < .6

- increased spread (angle) of jet

- faster decay of jet properties axially

- double hump velocity and conc. profiles

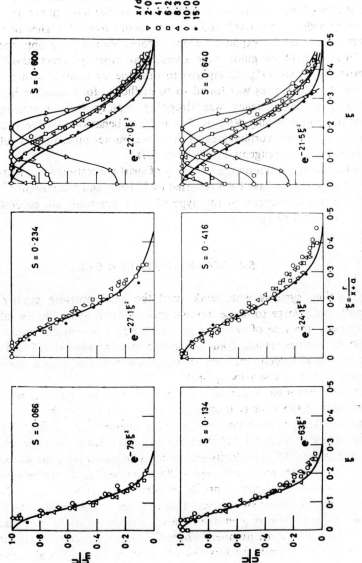

Fig. 5.10. Radial distributions of axial velocity (Chigier and Chervinsky[20]).

by extrapolating lines of $1/U_m$ against x to the value of $1/U_m = 0$, was found to be equal to 2·3 diameters, and independent of swirl.

Similarity axial velocity profiles satisfy the equation

$$U/U_m = \exp\left(-K_u r^2/(x + a)^2\right) \tag{2.3a}$$

The variation of the distribution constant with swirl is given by the empirical expression

$$K_u = \frac{92}{(1 + 6S)} \tag{5.20}$$

Radial distributions of the swirl velocity showed similarity for all the degrees of swirl investigated, from a distance of two diameters from the exit. The influence of the potential core is smaller on the w- than on the u-profile, thereby allowing similarity to be achieved closer to the orifice. The inner region of the swirl velocity profile is almost linear, corresponding to solid body rotation, while the outer region is a free vortex-type motion. The size of the solid body core increases gradually from $r/(x + a) = 0·1$ to 0·2 as swirl increases and then becomes about 0·1 again with higher swirl in which there is a recirculation zone. Similar swirl velocity profiles are described as third order polynomial equations with empirical constants. Gaussian error curves were also found to provide a good fit to measured profiles of static pressure.

For weak swirl the potential core extends to a distance of two diameters from the orifice and after four diameters the hyperbolic decay of maximum values of axial velocity prevails along the axis according to the equation

$$U_m/U_{m_o} = [6·8/(1 + 6·8S^2)]d/(x + a) \tag{5.21}$$

Decay of swirl velocity maxima was found to be independent of swirl in accord with the theoretical predictions which showed that the swirl velocity decay constant is equal to $(a/d)^2$.

The angle of spread of the jet, α, determined from the constant velocity line $U = 0·5U_m$, increases with S. These experimental results,[20] together with the observation of Gore and Ranz,[22] indicated that the angle of spread of the jet did not increase continuously with increase in S but, rather, tended to approach a certain asymptotic value. Kerr and Fraser,[21] using swirl vanes, obtained jets with swirl parameters up to 1·06 and their results indicate a linear increase of α with S. Examination of the different experimental conditions used by the various investigators shows that, if there is a wall at right angles to the flow direction and in the plane of the orifice, the angle of spread does not increase continuously with S. In the absence of such a wall, Gore and Ranz[22] found that, when the swirl was generated at the orifice (plate flush with orifice), the angle of spread increased continuously with S until, at the highest degree of swirl, the jet expanded at right angles and the whole region in front of the jet was in backflow. When the swirl was generated upstream from the orifice, as was the case with Gore and Ranz (plate recessed) and the tangential slot generator of Chigier and Chervinsky,[20] it was only possible to increase the angle of spread up to a limiting value of $\alpha = 10°$. For weak swirl the experiments showed

$$\alpha = 4·8 + 14S \qquad (5.22)$$

Mass flow rates within the jets were calculated by integration of the measured U-profiles. The entrained mass flow rate varies linearly with x according to the equation

$$M_e/M_0 = (0·32 + 0·8S)x/d \qquad (5.23)$$

Kerr and Fraser[21] studied two systems in which the swirl was imparted by swirl vanes. On one, a model, the swirl was computed from a direct measurement of the torque. In the second, an actual burner, there were two passages with swirl vanes, in some cases with counter-rotating swirl. In this case velocity measurements were used to determine the axial flux of angular momentum. Results compared favourably with those of other investigators.

Livesey et al.[13] measured the cold flow pressure drop through the diffuser of a half-scale model of an Admiralty burner. Two similar sizes were used, $3\frac{1}{2}$ in and $4\frac{1}{2}$ in in diameter, but both with bent swirl vanes set with an exit angle of $45°$. For all runs, the pressure drops decreased as the Reynolds number increased to about 800 000 and then started to increase again, but at a slow rate.

DIFFUSION FLAMES WITH WEAK SWIRL
Flames with weak swirl have only a limited practical interest mainly because of their tendency to instability. Under special conditions, weak swirl can also contribute to the lengthening of the flame which may be desirable in a particular application. Lack of recirculating flow allows the use of boundary layer approximations and improves the chances of theoretical treatment of this system.

Velocity and temperature measurements were made by Chigier and Chervinsky[23] in a series of turbulent, swirling-free flames with swirl numbers up to 0·214. Liquefied petroleum gas was injected into the tangential-slot swirl generator used previously in their isothermal studies. After ignition by a pilot flame, the flames stabilised some four diameters downstream from the burner exit where velocity magnitudes and velocity gradients were sufficiently reduced by entrainment to allow an annular flame front to be stabilised in the boundaries of the jet.

A relatively cold, high velocity core persists over almost the whole length of the flames, with the main reaction zone confined to the envelope between the cold central core and the surrounding air. No significant variation in the maximum temperature occurs along the length of the flame since, at each axial station, the maximum temperature is within the reaction zone. In the cold core, temperatures rise in the downstream direction largely due to turbulent mixing with hot combustion products from the reaction zone.

Radial distributions of axial velocity, U/U_m plotted against $r/(x + a)$, have similar forms and can be expressed in terms of a Gaussian equation. The velocity error curve constant, K_u, was found to increase with increase in swirl number. Normalised radial distributions of swirl velocity have the form of a combined vortex with an almost linear inner region corresponding to solid body rotation and an outer region corresponding to free vortex flow. The similar form of the normalised swirl velocity can be described as a function of $r/(x + a)$ by a third-order polynomial. Radial temperature distributions have temperature maxima in the main reaction zone and not on the jet axis. The temperature in the central core increases, due to mixing with hot combustion gases, and full similarity cannot be obtained under these conditions until farther downstream where the position of the temperature maximum converges on to the jet axis. In the outer region of the flow, temperature profiles have similar forms.

The decay of axial velocity increases with the degree of swirl. Decay constants are larger than in cold swirling jets, indicating a slower decay in axial velocity. The decay of swirl velocity does not show any significant variation with the degree of swirl as was found previously in cold jets. The swirl velocity decays in the flame as $1/x^2$ and the decay constant is considerably larger than in the cold swirling jet, again indicating a slower decay of swirl velocity in the flame.

5.5 STRONG SWIRL ($S > 0.6$)

When the swirl intensity is increased in a jet, a point is reached when the adverse pressure gradient along the jet axis cannot be further overcome by the kinetic energy of the fluid particles flowing in the axial direction, and a recirculating flow is set up in the central portion of the jet between two stagnation points. This recirculation zone, which has the form of a torroidal vortex, plays an important role in flame stabilisation, as it constitutes a well mixed zone of combustion products and acts as a storage of heat and of chemically active species located in the centre of the jet near the burner exit. In contrast to bluff body wakes, there is no need here for a solid surface that is exposed to high temperature and to the effect of deposition of coke.

Flame parameters, such as stability and combustion intensity, and residence time distributions in combustors depend on the size and strength of the vortex, and it is therefore of interest to determine these as functions of nozzle geometry and input flow conditions.

The Effect of the Degree of Swirl

As a result of swirl, the angle of spread of the jet increases with the swirl number. Corresponding to this increase in the spread of the jet, the entrainment increases causing faster decay of the velocity and nozzle fluid concentration with distance from the orifice. This is illustrated by Fig. 5.11 which represents the decays of the maximum axial, tangential and radial velocity components along the length of the jet and for various degrees of swirl.[6] Even though data in Fig. 5.11 are computed from measurements made in the region where the jets had not reached their fully developed

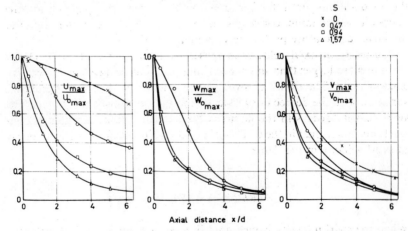

Fig. 5.11.   Decay of the maximum values of the axial, tangential and radial components of velocity along the length of the jets.

state, the decay of the velocities and pressure tend to the form predicted from theory,[24] *i.e.* the axial and radial velocity components decay as x^{-1} and the tangential component as x^{-2}. The pressure field decays even more steeply with distance from the orifice, as x^{-4}. Figure 5.12 shows the growth of the internal reverse flow zone with the degree of swirl. The

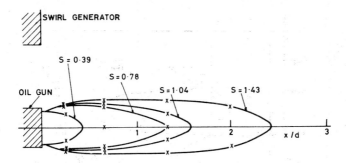

Fig. 5.12.   *Size of recirculation zone as a function of swirl number.*

boundaries of the reverse flow zones shown in Fig. 5.12 were determined in an annular system and a small reverse flow region is produced in the wake of the central bluff body even without swirling flow. For the case of $S = 0.39$ the size of the reverse flow zone is the same as that generated with zero swirl. It can be concluded that the transition swirl numbers determined for the round swirling jet also apply to annular swirling jets.

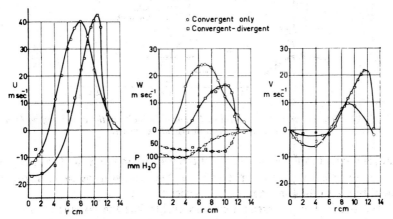

Fig. 5.13.   *Comparison of the velocity and static pressure distributions at a distance of $x = 10$ cm from the convergent and convergent–divergent nozzles.*

THE NOZZLE GEOMETRY

The shape of the exit nozzle, in particular the attachment of a divergent extension and blockage of the nozzle by either a cylindrical tube (*e.g.* pressure jet oil gun) or a stabiliser disc, will affect the flow pattern of a swirling jet. Figures 5.13 and 5.14 represent the radial distributions of the axial,

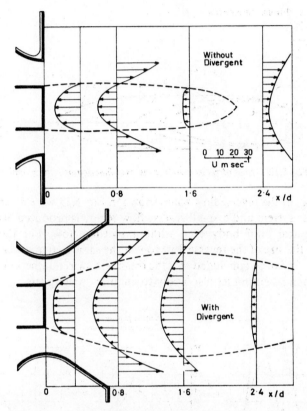

Fig. 5.14.   *Comparison of the radial distributions of velocity in the vortex region for jets with and without divergent extensions.*

tangential and radial components of the velocity in the jet and the position of the reverse flow zone respectively for two jets with the same degree of swirl but with different nozzle geometries.[2]

The use of the divergent nozzle extension has the combined effect of increasing the radial distance of separation between peaks of axial and tangential velocities and of also increasing the reverse mass flow rate.

Experiments in which the half angle and length of the divergent extension and the blockage of the nozzle were varied[6] showed that the
optimum half angle of the divergence is about 35° and its recommended
length $L = 1$ to $2d$, where d is the nozzle throat diameter.

The flow pattern in the reverse flow region is also affected by the
blockage caused by a stabiliser bluff body or by a fuel pipe or pressure jet
oil gun. At low swirl numbers ($S < 0.6$) the blockage is instrumental in
setting up a reverse flow region. As the degree of swirl is increased, the
effect of the blockage upon the size of the reverse flow region becomes

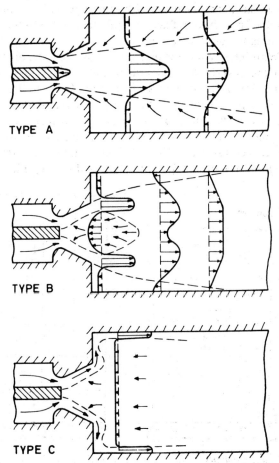

Fig. 5.15. *Flow forms observed for swirling jets and flames issuing from divergent nozzles.*

negligible, but at very high levels of swirl, or when a divergent nozzle is combined with swirling flow, a blockage of the nozzle throat may have an adverse affect upon the strength and size of the recirculating vortex in the centre of the jet.

SEPARATION OF SWIRLING FLOW IN DIVERGENT NOZZLES[36]
During investigations in which the interaction of nozzle geometry and swirling flow were studied,[17] three basic flow types were observed (Fig. 5.15).

Type A. With zero or low swirl intensity, the annular air flow tends to separate from the nozzle surface near the nozzle throat. This results in a jetting flow similar to that in an annular jet without swirl. Flames produced with this type of flow have a fluctuating flame front positioned some distance downstream from the burner exit.

Type B. With intermediate or high degree of swirl, the air flow is stably attached to the divergent nozzle walls and encloses a large torroidal vortex reverse flow zone situated in the centre of the jet. This is generally the desired flow pattern that produces highly stable flames with ignition close to the exit of the nozzle or even inside the divergence. This pattern also enables the matching of zones of high turbulence intensity with those of high fuel concentration, resulting in high combustion intensity.

Type C. This flow pattern represents the radial wall jet. Flames produced

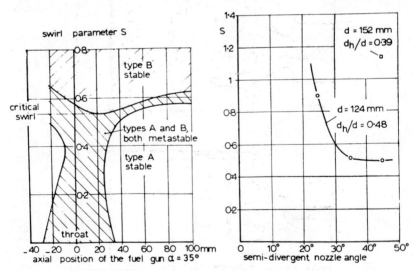

Fig. 5.16. Relative stability of flow forms A and B. Critical swirl as a function of semi-divergent angle.

with this flow type are attached to the burner face and to the wall of the combustion chamber (sunflower pattern). The burner geometry is critical for the establishment of radial wall jets. Trumpet-shaped wide angle and short divergent nozzles favour wall jet formation even at moderate swirl levels. The flames so produced are suitable in furnaces where high and uniform radiant heating of large heat sink areas are required.

The relative stability of flow between flow patterns[17] A and B is illustrated in Fig. 5.16 as a function of the swirl degree and the position of the blockage (oil gun) in the burner. The value of the critical swirl degree, the minimum swirl required for establishing type B flow pattern, is a function of the half angle of the divergent nozzle as shown in Fig. 5.16.

RESIDENCE TIME DISTRIBUTION

The effect of swirl on the performance and efficiency of a combustion chamber has been shown to be closely related to changes in the residence time distribution in the combustor due to variation in the degree of swirl. Reporting on a series of model and prototype studies carried out at Ijmuiden, Beér and Lee[25] showed that the combustor volume can be assumed to consist of a 'well stirred' part and of a 'plug flow' part in series with the stirred region. They also showed that the performance can be optimised for a highly loaded combustor by appropriately varying the proportions of the mean residence time spent in the well stirred and plug flow parts respectively.

Figure 5.17 shows results of helium tracer experiments carried out in pulverised coal flames in the Ijmuiden tunnel furnace compared with those obtained in a 1/10 scale water model. The graph represents the decay of tracer concentration in the exit stream after the tracer introduction to the burner has been cut off. The tracer concentration decay diagrams obtained in the furnace showed good agreement with those in the water model for cases in which the swirl number, corrected for isothermal modelling, was used as a modelling criterion:

$$S = \frac{G_\varphi}{G_x R_e} \qquad (5.12f)$$

$$R_e = R_0 \left(\frac{\rho_c}{\rho_h}\right)^{1/2} \qquad (2.11c)$$

where R_e is the equivalent nozzle radius. Figures 5.17 and 5.18 show that residence time distribution can effectively be controlled by swirl and that the proportion of residence time in the stirred section $t_s/\bar{t}$ as a function of

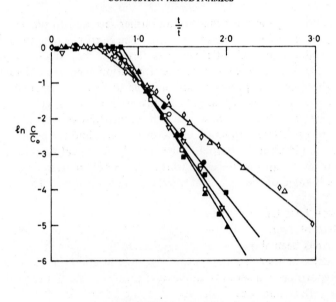

	$\dfrac{G_\phi}{G_x R}$		$\bar{t}$ sec.
◊	0	FURNACE.	15·2
△	0	MODEL.	20.
■	1·56	FURNACE.	18·3
▲	1·56	MODEL.	20
□	2·24	FURNACE.	17·9
▽	2·24	MODEL.	20
○	3·9	FURNACE.	19·3
●	3·9	MODEL.	20

Fig. 5.17. *Tracer concentration decay as a function of swirl number in model and prototype furnace.*

swirl number S goes through a minimum as the swirl degree is increased. Following these studies by the IFRF, Drake and Hubbard[11] showed that, for oil-fired boiler combustion chambers, the degree of swirl that gives the minimum $t_s/\bar{t}$ value yields the maximum performance with the lowest smoke emission.

Measurements of temperature distributions and calculations of burn out from the analysis of samples taken from pulverised anthracite flames have borne out the significance of swirling flow both in stabilising the flame and also in controlling the burn out of combustible material. Figure 5.19 shows

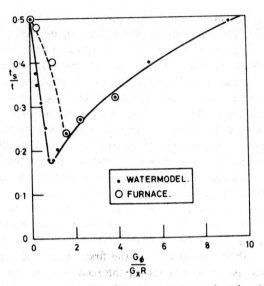

Fig. 5.18.   *Residence time in stirred section as a fraction of total residence time as a function of swirl number.*

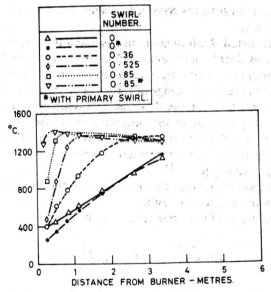

Fig. 5.19.   *Gas temperatures along the axis of pulverised-anthracite flames as a function of the degree of swirl.*

that the rate of temperature rise is greatly increased according to the degree of swirl. The flame front in these anthracite flames was generally observed in the region where the temperatures had risen to between 1000° and 1200°C. Without swirl the distance to the flame front was of the order of 3 m. This distance is reduced according to the degree of swirl.

The percentage of unburnt carbon determined from solid particle samples taken along the axis of the flames was shown, for anthracite flames, to decrease strongly as the degree of swirl was increased.[40]

More recently, the principle of varying residence time distribution by means of swirling flow was applied to the design of burners for a two-stage combustion process. By splitting the combustion air into a primary stream (15 to 20% of stoichiometric) and a secondary stream and by introducing the fuel (liquid or gaseous) with the strongly swirling primary air into a precombustion chamber with divergent nozzle exit, luminous flames could be produced even when burning natural gas. Good flame stability is ensured by the first stage of this process, while complete combustion is achieved in the second stage. Such a two-stage system is also suitable for reducing the emission of oxides of nitrogen from combustion processes.

TURBULENCE PROPERTIES IN STRONGLY SWIRLING JETS (WITH RECIRCULATION)

There is little detailed information on mean and fluctuating velocities and on shear stress distributions, even in constant density swirling flows. For prediction purposes, long-term interest has centred on numerical methods which in turn require knowledge of turbulence properties such as effective viscosity distributions in the flow field. Developments in measurement techniques, in particular in the interpretation of hot wire anemometer signals for very high intensity turbulence (see Chapter 8), enabled experimental studies to be made of recirculation zones set up in the central region of a strongly swirling jet.[37,38]

The relationships between the shear stresses and the velocity gradients also contain the effective viscosities.

The components of the stress tensor can be given as

$$\tau_{zz} = \rho\overline{U'^2} = -\mu_{\text{eff}(zz)}\left(2\frac{\partial\bar{U}}{\partial z}\right) \tag{5.24a}$$

$$\tau_{rr} = \rho\overline{V'^2} = -\mu_{\text{eff}(rr)}\left(2\frac{\partial\bar{V}}{\partial r}\right) \tag{5.24b}$$

$$\tau_{\theta\theta} = \rho \overline{W'^2} = -\mu_{\text{eff}(\theta\theta)} \left(\frac{\bar{V}}{r}\right) \tag{5.24c}$$

$$\tau_{zr} = \rho \overline{U'V'} = -\mu_{\text{eff}(zr)} \left(\frac{\partial \bar{U}}{\partial r} + \frac{\partial \bar{V}}{\partial z}\right) \tag{5.24d}$$

$$\tau_{z\theta} = \rho \overline{U'W'} = -\mu_{\text{eff}(z\theta)} \left(\frac{\partial \bar{W}}{\partial z}\right) \tag{5.24e}$$

$$\tau_{r\theta} = \rho \overline{V'W'} = -\mu_{\text{eff}(r\theta)} \left[r \frac{\partial}{\partial r} \left(\frac{\bar{W}}{r}\right)\right] \tag{5.24f}$$

Values of μ_{eff} can be derived from velocity gradients calculated from known mean flow field data and from measured shear stress distributions.

Spatial distributions of the stream function, determined in a strongly swirling flow by Syred et al.,[37] are shown in Fig. 5.20. The eye of the vortex is seen to be just outside the swirl generator (burner) exit. Entrainment of air from the surroundings is high with maximum values of $Q_e/Q_o = 1.8$ at $x/d = 0.5$.

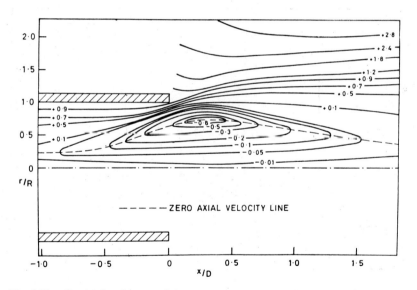

Fig. 5.20. *Spatial distributions of the stream function determined in a strongly swirling flow, $S = 2.2$.*

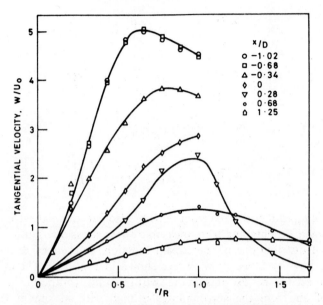

Fig. 5.21.   Radial distributions of tangential velocity.

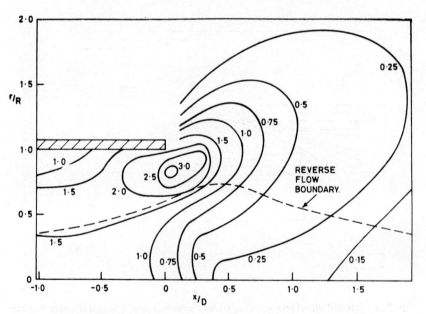

Fig. 5.22.   Spatial distributions of kinetic energy of turbulence k.

Radial distributions of swirl velocity are represented in Fig. 5.21.

Gradients of axial and swirl velocity components were found to be large, particularly in the radial direction (of the order of $2 \times 10^4 \text{ sec}^{-1}$). The positions of maximum velocity gradients coincide with those of the zero velocity line (dotted line in Fig. 5.20).

Distributions of kinetic energy of turbulence

$$k = \frac{(\overline{U'^2} + \overline{V'^2} + \overline{W'^2})}{U_0{}^2} \tag{5.25}$$

determined from detailed measurements of turbulence,[38] are shown in Fig. 5.22. Turbulence quantities have been normalised with the average exit velocity calculated as

$$U_0 = \frac{4Q_t}{\pi d^2} \tag{5.26}$$

Maximum turbulence can be seen to reach values of 300 % in the zone where flow separates from the swirl generator wall.

The spatial distributions of shear stresses, determined from measurements of all six turbulent stress components at a large number of points,[38]

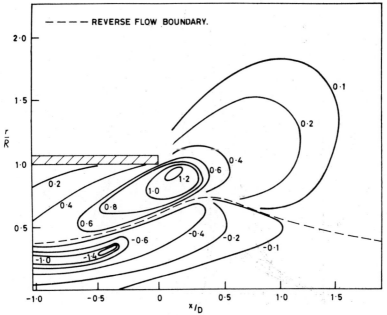

Fig. 5.23.   *Spatial distributions of normalised shear stress* $(\overline{W'U'}/\overline{U}_0{}^2)$.

showed that the most significant stress component in a flow with strong swirl is that in the '$z\theta$' plane (Fig. 5.23). Maximum stresses can be seen to occur close to the swirler exit and also in the recirculation zone well inside the swirl generator (zone of negative values in Fig. 5.23). Past the exit from the nozzle, values of the shear stress decay approximately twice as fast as do values of the kinetic energy of turbulence.

From measured turbulent stresses and the mean velocities, the various effective viscosities were computed. The most significant viscosities were found to be $\mu_{z\theta}$, $\mu_{\theta\theta}$ and μ_{zz}. In contrast, μ_{rr}, μ_{zr} and especially $\mu_{r\theta}$ were smaller by an order of magnitude.

Figure 5.24 represents the variation of $\mu_{\text{eff}(z\theta)}$, the viscosity of greatest importance, as a function of radial position at various axial distances in the '$z\theta$' plane. Figure 5.25 shows a comparison of five of the six viscosities' distributions.

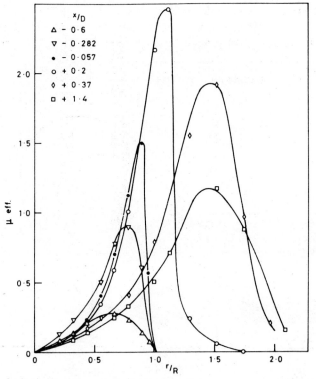

Fig. 5.24. Radial distributions of effective viscosity (WU direction).

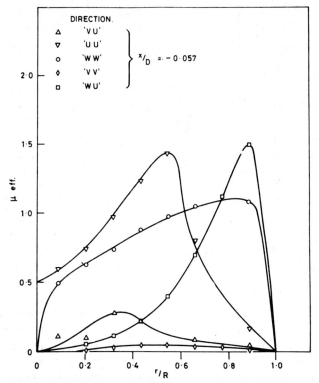

Fig. 5.25. Comparison of effective viscosity distributions.

The results in Figs. 5.24 and 5.25 show that values of effective viscosity are strongly dependent upon the position in the flow. Turbulence in swirling flows is strongly anisotropic.

It has been suggested[39] that a relationship of the form

$$\mu_{\text{eff}} = \rho k^{1/2} l C \tag{5.27}$$

may be used for describing effective viscosity distributions. It would appear from the above results that this formula needs to be modified for use in swirling flows. Possible solutions might be either to use different values of l and C for each of the six values of the effective viscosity or different values of directional velocity fluctuation for each of the six values of μ_{eff}, while maintaining common values of l and C in the computations. Much more detailed experimental and theoretical research is needed with the objective of establishing sufficiently general relationships between turbulence properties, their variation in the flow field on one hand and mean flow parameters on the other. This is necessary for the further development of prediction procedures in general and especially for their application to strongly swirling flows.

5.6 BUOYANT OR TURBULENT JET DIFFUSION FLAME IN A ROTATING FLOW ENVIRONMENT

The typical system of this type consists of a buoyant column of gas surrounded by a rotating environment. Emmons,[28] who has studied the fire whirl from a liquid fuel pool formed at the centre of a rotating cylindrical screen, gave a clear description of this type of flow. The phenomenon of fire whirl is produced by a concentrating mechanism which brings vorticity together into the vortex core of the rotating system. In the case of flames, the concentrating process is the rising gas in the buoyant column. In addition to this, the boundary between the flame and the surrounding air is stable because the centrifugal force opposes entrainment of the air into the rising column once it is above ground level. Vorticity is concentrated into the core, where again the fluid is stable with regard to radial interchanges of fluid as shown by Rayleigh[29] (rotating fluid is stable if $\rho w r$ increases with r). As an extension of the Emmons investigation, an experimental study was carried out at Sheffield on a system where the pool burning in the centre of the rotating screen was replaced by a turbulent jet.[5]

In diffusion flames, flame length is essentially dependent upon the rate of mixing between fuel and air. Entrainment of air into the core is reduced and, since fuel is introduced and remains within the vortex core, the consequent delay in mixing results in increased flame length. Flame lengths for methane diffusion flames in the range of Reynolds numbers 400 to 5000 are shown in Fig. 5.26 for a burner exit diameter, $d = 3\cdot17$ mm, where flame length/nozzle diameter is plotted as a function of Reynolds number based on nozzle exit conditions. Flame length measurements for zero circulation were found to lie between the values calculated from the empirical formulas of Hottel and Hawthorne[30] and Wohl et al.[31] These flames were laminar in the initial region near the nozzle exit, followed by a turbulent brush. Increases in flame length are mainly due to the effect of rotation upon the turbulent brush of the flame.

The stability of the boundary layer at the interface between the vortex core and the free vortex region is affected by the fact that this interface is also the boundary of the cylindrical flame. Coupled with the centrifugal force field, the radial density gradient in the flow causes stratification and the boundary to become even more stable. This effect is illustrated by the Schlieren photographs which show the flame without and with the rotation of the screen (Fig. 5.27).

Subsequent experiments[32] were carried out in a vertical, stationary, cylindrical tube mounted on a variable block swirl generator with a central, burning, gaseous fuel jet. Measurements were also made in an

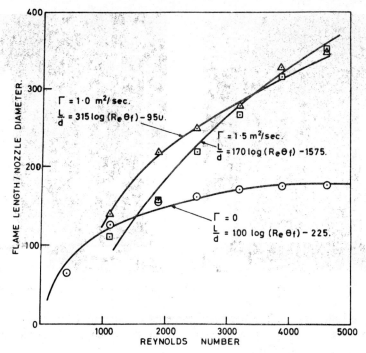

Fig. 5.26. *Flame lengths for methane diffusion flames in a rotating atmosphere,* $d = 3\cdot17$ mm.

isothermal model of the system with a helium jet replacing the fuel jet.

In the isothermal model, helium concentration was measured with the gas phase chromatograph and mean and fluctuating velocities were measured with a DISA miniature hot wire anemometer probe. The anemometer readings were interpreted using a method[33] which allowed the determination of all the mean velocity components and the turbulence stresses by placing the hot wire in six different positions at each measuring point.

Photographs of the flame in the tube under conditions of full rotation showed the flame front to be of helical form with burning confined to the outer edge of the rotating core. The flame was cylindrical in shape with smooth edges, in contrast to the jet flame without rotation, spreading out radially with turbulent eddies visible on the edges.

Isotherms within the flame show elongation of the flame together with deflection of the high temperature reaction zone away from the axis in the

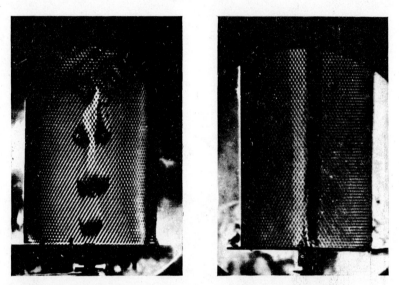

Fig. 5.27.   Schlieren photographs of flames; left, stationary screen; right, rotating screen.

case of full rotation. Temperature gradients at the outer edges of the flame were higher with full rotation while visible flame lengths were increased from $L/d = 180$ to 320.

With full rotation, flames were highly luminous with large quantities of soot, and measured oxygen concentrations were lower within the flame. These temperature and concentration measurements are further evidence of the reduced entrainment of air into the flame.

Isothermal Model

MEAN VELOCITY AND DENSITY
Velocity measurements were made in the vortex tube with the helium jet using the DISA hot wire anemometer. Profiles of the tangential mean velocity components at x/d between 4 and 80 showed no significant change

in the velocities of the air outside the helium jet. The tangential velocity profile had the form of a Rankine vortex with solid body rotation in the inner two-thirds of the radius. Measurement of these profiles without the presence of the helium showed no significant difference.

Density distributions were calculated from measurements of gas concentrations using the gas phase chromatograph. Density variations, due only to changes in helium concentration, are shown in Fig. 5.28. The densities are seen to increase as a result of mixing with the surrounding air.

Previous studies in the rotating screen had shown the importance of having a buoyant jet in order to provide a 'sink' for the vortex. The helium jet was chosen to model the methane flame at a temperature of 1400°C.

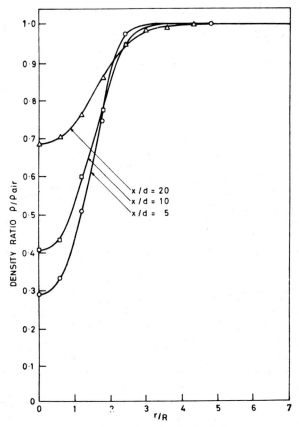

Fig. 5.28. *Radial distributions of normalised density for helium jet in a rotating, co-axial air stream.*

THE EFFECTS OF A STRONG, POSITIVE, RADIAL DENSITY GRADIENT COUPLED
WITH ROTATION

The effects of changes in jet density and rotation of the surroundings were
first investigated separately and then the combined effects were subsequently studied. Figure 5.29 represents a comparison between an air jet
and a helium jet in a non-rotating, co-axial air stream at $x/d = 20$. Mean

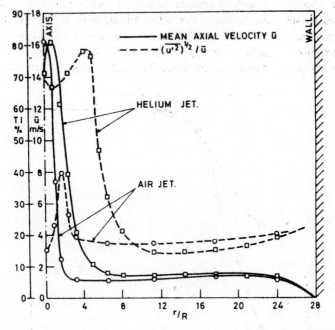

*Fig. 5.29.   Effect of jet density on the radial distribution of turbulence intensity $(\overline{u'^2})^{1/2}/\bar{u}$ in
a jet in a co-axial, non-rotating air stream at $x/d = 20$.*

axial velocity components and axial turbulence intensity, $(\overline{u'^2})^{1/2}/\bar{u}$, are
shown for both cases. Turbulence intensities and spread of the helium jet
are twice as large as in the case of the corresponding iso-density air jet
system. These results are directly due to the decrease of the jet density.

Figure 5.30 shows the effect of rotation of the surroundings on the isodensity air jet system. Rotation alone causes a marked decrease in the shear
stresses within the jet. The shear stresses generated within the rotating flow
are well away from the mixing region between jet fluid and surroundings
and play no part in mixing near the jet boundary.

Figure 5.31 shows the combined effect of rotation and density on the shear stress components for the case of a helium jet introduced into the co-axial air stream: the shear stresses are drastically reduced within the jet. Whereas density differences normally result in increased mixing, the combined effect of rotation and of a positive radial density gradient has resulted in a greater decrease in mixing than that caused by rotation alone.

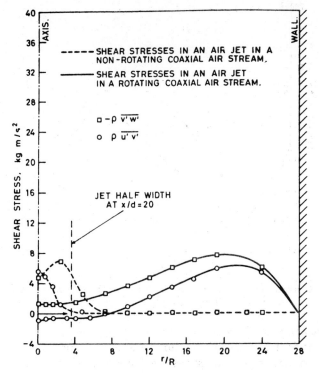

Fig. 5.30. *Effect of rotation of the surroundings on the shear stress components in an air jet, x/d* = 20.

In the combined case, the helium jet acts as a buoyant column in which vorticity is concentrated and where the boundary between the helium jet and the surrounding air is stable.

The results of turbulence characteristics obtained in the isothermal model substantiate the measurements of temperature and the photography of the flames. The laminarisation observed initially in the rotating screen system and subsequently in the vortex tube can be explained in terms of the stability imposed on the system as a result of rotation.

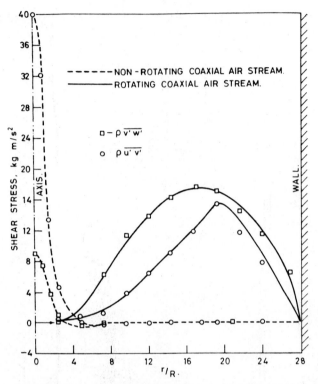

Fig. 5.31.   *Combined effect of rotation and jet density on the shear stress components for a helium jet in a rotating, co-axial air stream, $x/d = 20$.*

LAMINARISATION OF FLAMES

The transition from laminar to turbulent flames normally occurs at a critical Reynolds number when the break point starts at the flame tip and rapidly moves down towards the burner with increasing velocity. For methane flames, critical Reynolds numbers are between 3000 and 4000 and both direct and Schlieren photographs show the turbulent brush with large eddies clearly visible at the flame boundaries.

The imposition of an appropriate external force field has been shown to have a stabilising effect upon the transition from laminar to turbulent flow.[34] In a centrifugal force field, such as flow in the annulus between two rotating cylinders, a fluid particle from an outer layer opposes a tendency to being moved inwards because its centrifugal force exceeds that acting on a particle near the axis of the cylinder and thus it shows a tendency to be thrown outwards. Equally, motion outwards is against the force field since the centrifugal force acting on an inner particle is smaller than that on a particle further away from the axis. Transverse motions are thus impeded by the stabilising effect of the centrifugal forces. The system is analogous to the influence of vertical density variations on the stability of horizontal flow. Phenomena occur in meteorological processes where stable density stratification impedes turbulent mixing and suppresses turbulence.

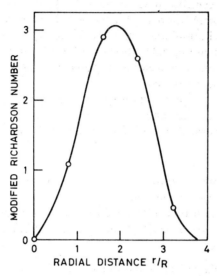

Fig. 5.32. Radial distribution of modified Richardson number at x/d = 80.

Prandtl[35] used an energy method to analyse systems with density gradients as well as those under the influence of centrifugal forces. He showed that the stability of stratified flows depends upon the stratification parameter

$$\text{Ri} = - \frac{(g/\rho)(\mathrm{d}\rho/\mathrm{d}y)}{(\mathrm{d}U/\mathrm{d}y)_w{}^2} \qquad (5.28)$$

known as the Richardson number, in addition to the usual dependence on Reynolds number. The subscript w refers to the value of the velocity gradient at the wall.

For the case of a flame in a rotating air flow, the gradients of angular momentum are coupled with density gradients. Under stabilising conditions these forces act against the shear forces which generate turbulence. The shear forces are directly proportional to the velocity gradients $\mathrm{d}U/\mathrm{d}r$. If we consider gravitational forces to be of secondary importance and take the laminarisation parameter to be the ratio of centrifugal forces in a field to density gradients to the shear forces, we obtain[37]

$$Ri^* = \frac{(1/\rho)(\partial\rho/\partial r)(W^2/r)}{(\partial U/\partial r)^2} \tag{5.28a}$$

In this modified Richardson number, gravitational acceleration has been replaced by centrifugal acceleration.

Values of Ri^* were computed for the helium jet in the rotating air flow in which laminarisation was established, and the radial distribution at $x/d = 80$ is shown in Fig. 5.32. Maximum values of $Ri^* = 3$ were found in the region of maximum concentration gradients. This is the region in which the flame envelope is situated.

Stabilising effects begin to act for $Ri^* > 0$, but for $Ri^* > 1$ the stabilising forces become dominant and succeed in damping the turbulence in the system as demonstrated in both the flame and helium jet experiments.[32]

CONCLUSIONS

Experimental work in three systems, (i) rotating screen, (ii) flame in vortex tube, and (iii) helium jet in vortex tube, has in each case shown laminarisation of a turbulent jet as a direct result of rotation in the presence of a density field. As a result of the rotating environment the following effects were found:

1. increase in flame length and in flame luminosity;
2. decrease in rate of spread, oxygen concentration, turbulence intensities and shear stress in the flame;
3. rapid initial expansion of the flame due to high turbulence intensity and mixing rates near the nozzle exit resulting in improved blow-off stability of flames;
4. laminarisation of the boundaries of the jet as observed on direct and Schlieren photographs of the flames and damping of turbulence as measured by hot wire anemometer in the helium jet model;
5. stabilising effects begin to act for modified Richardson number $Ri^* > 0$, but for $Ri^* > 1$ the stabilising forces become dominant and succeed in damping turbulence in the system.

It can be concluded that the modified Richardson number is a valid criterion for determination of stability in flow systems with rotation and density gradients.

REFERENCES

1. Beér, J. M. and Chigier, N. A., 'Swirling jet flames from an annular burner', 5 me Journée d'Études sur les Flammes, Paris, 1963; also Doc. No. K20/a91, International Flame Research Foundation, Ijmuiden, Holland, 1963.
2. Chigier, N. A. and Beér, J. M., 'Velocity and static pressure distributions in swirling air jets issuing from annular and divergent nozzles', Trans. ASME, 86D, J. Basic Eng. 1964, 4, pp. 788–96.
3. Morton, B. R., 'Geophysical vortices', Prog. Aeron. Sci. 1966, 7, pp. 145–94.
4. Emmons, H. W. and Shuh-Jing Ying, 'The fire whirl', Eleventh Symposium on Combustion, The Combustion Institute, 1967, pp. 475–88.
5. Chigier, N. A., Beér, J. M., Grecov, D. and Bassindale, K., 'Jet flames in rotating flow fields', Combustion and Flame 1970, 14, pp. 171–9.
6. Chedaille, J., Leuckel, W. and Chesters, A. K., J. Inst. Fuel 1966, 39, No. 311, pp. 506–21.
7. Coursimault, J., 'Problèmes posés par l'utilisation du mazout dans les grandes chaudières', Ref. Gen. Therm. 1967, No. 65, pp. 617–33.
8. Sagalova, S. L., et al., 'Aerodynamics of the furnace space and development of the anthracite combustion process when operating with direct flow and turbulent burners', Thermal Eng. 1966, No. 7, pp. 49–55.
9. Sagalova, S. L., et al., 'Results of testing 10t/h p.f. burners operating on anthracite', Thermal Eng. 1967, No. 1, pp. 16–20.
10. Whitney, G. C., 'Use of models for studying pulverised coal burner performance', Trans. ASME Ser. A, Oct. 1959, pp. 380–2.
11. Drake, P. F. and Hubbard, E. H., 'Combustion system aerodynamics and their effect on the burning of heavy fuel oil', J. Inst. Fuel 1966, 39, No. 302, pp. 98–109.
12. Leikert, K., Konstruktive Merkmale von Kohlenstaubfeuerungen für grosse Dampferzeuger, VGB—Fachtagung 'Kohlenfeuerungen', 1968, pp. 4–23.
13. Livesey, J. L., Wilcox, P. L. and South, R. D., 'Isothermal aerodynamic investigation of a suspended air register', J. Inst. Fuel 1968, 41, No. 327, pp. 169-86.
14. Haykluytt, J. P. D. and North, B. C., 'The design of air registers for oil-fired boilers', Trans. Inst. Marine Eng. 1967, 79, No. 3, pp. 57–83.
15. Sagalova, S. L., et al., 'Aerodynamic characteristics of flow produced by a burner with guide vanes', Thermal Eng. 1965, No. 6, pp. 31–7.
16. Tissandier, G., 'Bruleurs à charbon pulvérisé à turbulence pour chaudières de centrales thermiques', Doc. No. G 13/a/2, International Flame Research Foundation, Ijmuiden, Holland, 1969.
17. Leuckel, W., 'Swirl intensities, swirl types and energy losses of different swirl generating devices', Doc. No. GO2/a/16, International Flame Research Foundation, Ijmuiden, Holland, 1968.
18. Kerr, N. M. and Fraser, D., 'Swirl—effect on axisymmetrical turbulent jets', J. Inst. Fuel 1965, 38, No. 299, pp. 519–26.
19. Mathur, M. L. and Maccallum, N. R. L., 'Swirling air jets issuing from vane swirlers', J. Inst. Fuel 1967, 40, No. 316, pp. 214–25.
20. Chigier, N. A. and Chervinsky, A., 'Experimental investigation of swirling vortex motions in jets', 5th US Nat. Congress on Applied Mechanics, Univ.

Minnesota, June, 1966; *Trans. ASME*, Ser. E, *J. Appl. Mech.* 1967, **34,** pp. 443–451.

21. Kerr, N. M. and Fraser, D., 'Swirl part 1, effect on axisymmetrical turbulent jets; part II, effect on flame performance and modelling of swirling flames', *J. Inst. Fuel* 1965, **38,** No. 299, pp. 519–38.

22. Gore, R. W. and Ranz, W. E., 'Back flows in rotating fluids moving axially through expanding cross sections', *Am. Inst. Chem. Engrs J.* 1964, **10,** No. 1, pp. 83–8.

23. Chigier, N. A. and Chervinsky, A., Eleventh Symposium on Combustion, pp. 489–99, The Combustion Institute, 1967.

24. Loitsyanski, L. G., 'The propagation of a twisted jet in an unbounded space filled with the same fluid', *Prikladnaya Matematika i Mekhanika* 1953, **17,** pp. 3–16.

25. Beér, J. M. and Lee, K. B., Tenth Symposium on Combustion, pp. 1187–1202, The Combustion Institute, 1965.

26. Beér, J. M., *VDI Berichte* 1970, No. 146, pp. 15–24.

27. Lilley, D. and Chigier, N. A., 'Non-isotropic turbulent stress distribution in swirling flows', *Int. J. Heat Mass Transfer* 1971, **14,** pp. 573–85.

28. Emmons, H. W. and Shuh-Jing Ying, Eleventh Symposium on Combustion, pp. 475–88, The Combustion Institute, 1967.

29. Lord Rayleigh, *Proc. Roy. Soc.* 1916, **A93,** pp. 148–54.

30. Hottel, H. C. and Hawthorne, W. R., Third Symposium on Combustion, pp. 254–66, Williams and Wilkins, Baltimore, 1951.

31. Wohl, K., Gazeley, C. and Kapp, N., Third Symposium on Combustion, p. 288, Williams and Wilkins, Baltimore, 1951.

32. Beér, J. M., Chigier, N., Davies, T. W. and Bassindale, K., 'Laminarisation of turbulent flames in rotating environments', *Combustion and Flame* 1971, **16,** pp. 39–45.

33. Siddall, R. G. and Davies, T. W., International Seminar on Heat and Mass Transfer, Herceg Novi, Yugoslavia, Sept. 1969.

34. Schlichting, H., *Boundary Layer Theory*, p. 492, McGraw-Hill, New York, 1968.

35. Prandtl, L., *Collected Works*, Vol. 2, p. 788, Springer-Verlag, Berlin, 1961.

36. Beér, J. M. and Leuckel, W., 'Turbulent flow in rotating flow systems', Paper No. 7, North American Fuels Conference, Ottawa, Canada, May 1970, Canadian Combustion Institute–ASME–Inst. Fuel.

37. Syred, N., Chigier, N. and Beér, J. M., Thirteenth Symposium on Combustion, The Combustion Institute, 1971, pp. 563–70.

38. Syred, N., Beér, J. M. and Chigier, N., 'Turbulence measurements in swirling recirculating flows', Institute of Mechanical Engineers, London. Symposium on Internal Flow, University of Salford, April 1971, paper 13, B27–36.

39. Gossman, A. D., Pun, W. M., Runchal, A. K., Spalding, D. B. and Wolfstein, M., *Heat and Mass Transfer in Recirculating Flows*, Academic Press, London and New York, 1969.

40. Beér, J. M. and Chigier, N. A., 'Stability and combustion intensity of pulverised coal flames—effect of swirl and impingement', *J. Inst. Fuel* 1969, **42,** pp. 443–50.

CHAPTER 6

Droplets and Sprays

NOMENCLATURE

a'	constant.
b'	constant.
B	transfer number [eqn. (6.17)].
c	specific heat of vapour–inert gas mixture.
C_D	drag coefficient.
$\overline{C_D}$	drag coefficient in presence of mass transfer.
d	droplet diameter.
d_0	initial droplet diameter.
d'_1, d'_2	diameter of thin spherical shells.
d_c	cenosphere diameter.
d_f	flame diameter.
d_∞	diameter of a spherical shell at infinite distance from the drop surface.
d_j	jet diameter.
d_l	ligament diameter.
D	mean diffusivity of gas mixture.
D_O	mean diffusivity of oxygen beyond the flame.
D_i	diffusion coefficient of species i.
D_m	swirl chamber diameter.
D_2	orifice diameter.
E	activation energy.
E'	energy.
f'	frequency.
g	acceleration due to gravity.
Gr	Grashoff number.
H	heat of combustion.
H_i	enthalpy of species i.
h	half sheet thickness.
h^*	half sheet thickness at break up.
i	stoichiometric ratio.

k	thermal conductivity at temperature T (vapour–inert gas mixture).
k_l	thermal conductivity of liquid.
k_i	thermal conductivity (vapour–inert gas mixture) at T_s [eqn. (6.11b)].
$\bar{k}$	mean thermal conductivity.
K	wavenumber $(= 2\pi/\lambda')$.
K_1, K_2, K_3, K_4	constants.
K'_1, K'_2	constants.
L_o	orifice length.
Le	Lewis number.
$\dot{m}$	mass burning rate.
$\dfrac{\mathrm{d}m}{\mathrm{d}t}$	mass burning rate.
$\dot{M}_s$	mass burning rate or mass rate of evaporation in absence of convection.
$\dot{M}_N$	mass burning rate or mass rate of evaporation under natural convection conditions.
$\dot{M}_F$	mass burning rate or mass rate of evaporation under forced convection conditions.
M_{O_2}	mass fraction of oxygen in the ambient gas.
$\overline{M_i}$	mean molecular weight of species i.
N_∞	mole fraction of oxygen in the bulk surrounding gas.
Nu	Nusselt number.
$\overline{\mathrm{Nu}}$	Nusselt number in presence of mass transfer.
P	pressure.
ΔP	differential ejection pressure in kg/sec^2 m.
p_1	amplitude of air vibration.
Pr	Prandtl number.
Q	heat required to evaporate unit mass of fuel.
Q_e	volumetric flow rate.
r	radial co-ordinate.
r_l	droplet radius.
r_f	flame radius.
R	universal gas constant.
Re	Reynolds number.
S	droplet spacing.
Sc	Schmidt number.
t	time.

t_b	burning time.
T_f	flame temperature.
T_s	surface temperature of the drop.
T_l	temperature of the liquid drop.
T_G	temperature of the ambient gas.
T	absolute temperature.
U	mean axial velocity.
V	air velocity, droplet velocity.
v	radial velocity.
X	number of moles of combustion products derived from stoichiometric combustion of unit mass of fuel with oxygen.
Y_i	weight fraction of species i.
Y_F	weight fraction of fuel vapour at the droplet surface.
Y_{O_∞}	weight fraction of oxygen at infinity.
Y_{O_i}	mass fraction of oxidiser in the ambient gas.
Y_{O_l}	mass fraction of the oxidiser at the surface of the drop.

Greek Symbols

α	thermal diffusivity of the liquid.
α_1	thermal diffusivity of the gas mixture.
γ	surface tension.
ρ	mean density of gas mixture.
ρ_L	density of liquid
ρ_l	density of the liquid drop at drop temperature (b.p.)
ρ_g	density of gas.
ρ_O	density of oxidant.
τ	flame propagation time.
η	viscosity of liquid.
μ	dynamic viscosity.
λ	burning or evaporation constant.
λ'	wavelength.
λ'_{opt}	optimum wavelength for jet disruption.
λ_s	burning constant with no vibrations.
λ_z	burning constant with vibrations.
Ω	number of moles of oxygen required for stoichiometric combustion of unit mass of fuel.

6.1 INTRODUCTION

The injection of liquid fuel into combustion chambers through atomisers facilitates the disintegration of the liquid into a spray of droplets. Most practical liquid fuel sprays have a size distribution over a wide range of drop sizes with a mean drop size between 75 and 130 μm and a maximum drop size preferably under 250 μm. The smallest drops vaporise completely but, with heavier fuels, liquid phase cracking takes place in the larger drops leading to the formation of carbonaceous residue often in the form of a cenosphere. When the fuel vapour is ignited in the vicinity of the droplets this results in an increased rate of evaporation. Local air/fuel ratios are determined by mixing between fuel in vapour or solid phase, combustion air and recirculated combustion products. Reaction zones will be formed partly surrounding each drop or in regions where air/fuel ratios are within the limits of inflammability. The rate of combustion of a spray is a function of the size and velocity distribution of drops, the properties of the fuel and its gaseous environment and the mechanics of the mixing processes.

Practical oil flames may be considered as being one of two basic types:

(1) Turbulent jet diffusion flames in which the oil is atomised by high pressure air or steam (blast atomised) and where the momentum of the fuel spray is so high that it is fully sufficient for the entrainment of combustion air necessary to complete the combustion. The significant dimensions of the flame, such as length and angle of spread, can be calculated from turbulent jet theory based on the fuel-atomising agent spray as the momentum source.

(2) Pressure jet flames in which the momentum of the spray is small in comparison with the momentum of the air flow. In this case the characteristic dimensions of the flame will depend more on the air flow pattern than on the fuel spray.

Most practical flames fall within these two categories. Solution of problems concerning flame stabilisation, rate of combustion, formation of carbon, and radiation properties of flames requires detailed knowledge of trajectories of droplets and rates of burning of drops, together with a statistical description of the drops in the spray with regard to size and spatial distributions. Other important practical problems such as carbon formation and deposition on combustion chamber walls are affected by both spray characteristics and air flow patterns.

6.2 ATOMISATION OF LIQUID FUELS

Liquid fuels are injected into combustion chambers by means of atomisers which cause the liquid to be disintegrated into drops lying within a specified size range and which control the spatial distribution of the drops. Atomisers are classified according to the basic forms of energy employed, namely, liquid pressure, centrifugal, kinetic energy of air or steam and ultrasonic energy. The most effective way of utilising energy imparted to a liquid is to arrange that the liquid mass has a large specific surface before it commences to break down into drops. Thus the primary function of an atomiser is to transpose bulk liquid into thin liquid sheets.

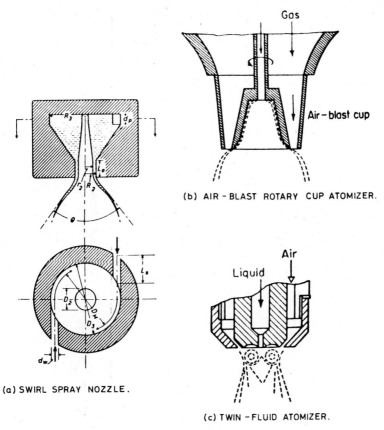

(b) AIR - BLAST ROTARY CUP ATOMIZER.

(a) SWIRL SPRAY NOZZLE.

(c) TWIN – FLUID ATOMIZER.

Fig. 6.1.   *Types of atomiser (after Dombrowski*[1]*).*

TABLE 6.1

Spray Characteristics
(after Dombrowski[1])

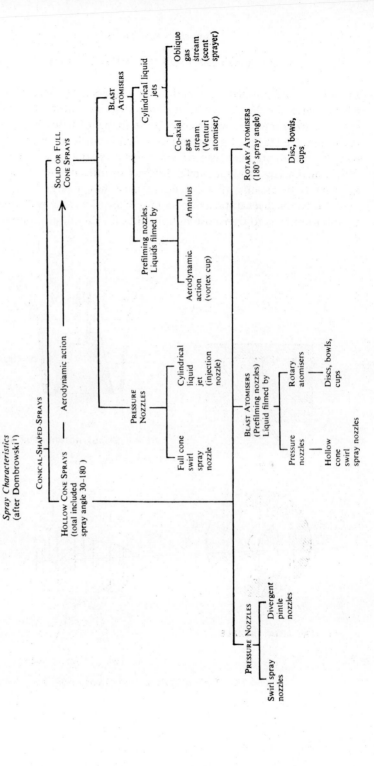

Basic spray characteristics are given in Table 6.1 which distinguishes between hollow cone sprays, in which the spray drops are concentrated in the periphery of a cone, and solid cone sprays, in which the whole volume of the spray cone is filled with an almost evenly distributed mass of drops. One of the most widely used atomisers is the pressure jet atomiser in which liquid is forced under pressure through a nozzle of the type shown in Fig. 6.1(a). Liquid enters the nozzle tangentially and emerges from the orifice in the form of a hollow conical sheet with an air core. Spray characteristics depend upon liquid pressure, liquid viscosity and nozzle dimensions such as the ratio of swirl chamber diameter to the orifice diameter (D_m/D_2) and the ratio of orifice length to orifice diameter (L_o/D_2). Effects of variation of pressure, fuel viscosity, ambient density, and correlations of discharge coefficients and spray angles, are given in a review on atomisation by Dombrowski.[1] From the equation of conservation of energy it follows that the flow rate varies with the square root of the pressure drop and, because of this relationship, pressure nozzles are somewhat inflexible since a large range of flow rates requires excessive variations in differential pressure. These limitations have been overcome in swirl spray nozzles by the development of spill, duplex, variable port and multiple orifice atomisers, in which ratios of maximum to minimum outputs in excess of 50% can be easily achieved.

In a rotary atomiser, liquid is fed onto a rotating surface and spread out by centrifugal force. Under normal operating conditions the liquid extends from the periphery in the form of a thin sheet which breaks down some distance away, either freely by aerodynamic action or by the action of an additional gas blast. Since the accelerating force can be independently controlled, this type of atomiser is extremely versatile. An air-blast rotary cup atomiser is shown in Fig. 6.1(b) in which the liquid film is formed on the inside surface of a rotating cup and the liquid sheet is disintegrated by a stream of pressurised air or steam.

In twin-fluid or blast atomisers, a high velocity gas stream impinges on a relatively low velocity liquid stream, either internally within the atomiser body or externally as shown in Fig. 6.1(c). When the required liquid flow rate is high, the liquid jet diameter must be increased, and under these conditions the energy transfer becomes very inefficient. Fraser[2] has studied air flow patterns outside the atomiser, but in general the design of twin fluid atomisers follows empirical lines since little is known of the behaviour of gas streams as they impact on liquid surfaces.

Less than 0·5% of the applied energy is utilised in breaking up liquids into small drops and, for pressure and twin fluid atomisers, practically the whole amount is imparted to the liquid or gas as kinetic energy. The rate of work required for pressure nozzles is given by

$$E' = 1\cdot331 \times 10^3 \, Q_e \, \Delta P \, \text{kg m}^2/\text{sec}^3 \qquad (6.1)$$

where Q_e is in m^3/sec and ΔP in kg/sec^2 m.

6.3   DROP FORMATION IN SPRAYS

The fundamental principle of the disintegration of a liquid consists of increasing its surface until it becomes unstable and disintegrates. The process by which drops are produced from a liquid stream depends upon the nature of the flow in the atomiser, i.e. whether laminar or turbulent, the way in which energy is imparted to the liquid, the physical properties of the liquid and the properties of the ambient atmosphere. The basic mechanism, however, is unaffected by these variables and consists essentially of the breaking down of unstable threads of liquid into rows of drops conforming to the classical mechanism postulated by Lord Rayleigh.[3] The theory states that a free column of liquid is unstable if its length is greater than its circumference, and that for a non-viscous liquid the wavelength of that disturbance which will grow most rapidly in amplitude is 4·5 times the diameter. Weber[4] has shown that, for viscous liquids, the optimum wavelength for jet disruption is

$$\sqrt{2}\pi d_j[1 + 3\eta(\rho_L\gamma d_j)^{-1/2}]^{1/2} \qquad (6.2)$$

A uniform thread will break down into a series of drops of uniform diameter, each separated by one or more satellite drops. Because of the

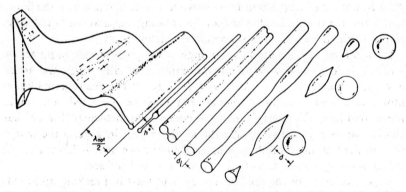

Fig. 6.2.   *Idealised process of drop formation from a liquid sheet (after Dombrowski[1]).*

irregular character of the atomisation process, non-uniform threads are produced which results in a wide range of drop sizes. An homogeneous spray can be produced only when the formation and disintegration of threads are controlled. The three modes of disintegration, namely, rim, wave and perforated sheet, are discussed in detail by Dombrowski.[1]

The complex process of drop formation from a sheet subject to aerodynamic sinuous waves has been idealised by Dombrowski[1] in Fig. 6.2. This shows that the waves on the sheet continue to grow until the crests are blown out. The sheet is thus broken up into half wavelengths which rapidly contract into ligaments which in turn break up into drops. The volume of liquid per unit width in a half wavelength of sheet is

$$\lambda'_{opt} h^* = \frac{\pi d_l^2}{4} \tag{6.3}$$

and thus the ligament diameter

$$d_l = 1 \cdot 128 \ \sqrt{(\lambda'_{opt} h^*)} \tag{6.4}$$

It has been found that ligaments produced from a low viscosity spray sheet at atmospheric gas density and low ejection pressures break down in accordance with Rayleigh's theory for the instability of cylindrical liquid columns. Analysis shows that, when the product of the wavenumber K and the half sheet thickness h, $Kh < 0 \cdot 25$, the drop size is a direct function of the surface tension γ and the liquid density ρ_L, and an inverse function of the liquid ejection pressure P and the gas density ρ_g. The equations also show that, when $Kh > 0 \cdot 25$, drop size is independent of surface tension or ejection pressure and that it increases with increasing gas density. For pressure nozzles, the flow rate is proportional to the square root of the pressure. The functional relationship between the operating variables and the dimensions and velocity of the sheet for swirl spray nozzles is unknown and drop size expressions for these atomisers are essentially based on empirical correlations.

A spray is a dynamic system whose quality may vary with time. The acceleration of a drop is an inverse function of its diameter and thus the spectrum of a polydisperse spray will change with distance from the nozzle, not only as a result of evaporation, but also according to the initial direction of the spray and the nature of imposed or induced air currents. Furthermore, coalescence of drops may take place, particularly if the ambient atmosphere is in highly turbulent motion. Methods of measuring drop size have been classified by Dombrowski[1] in Table 6.2.

TABLE 6.2

Methods of Measuring Drop Size
(after Dombrowski[1])

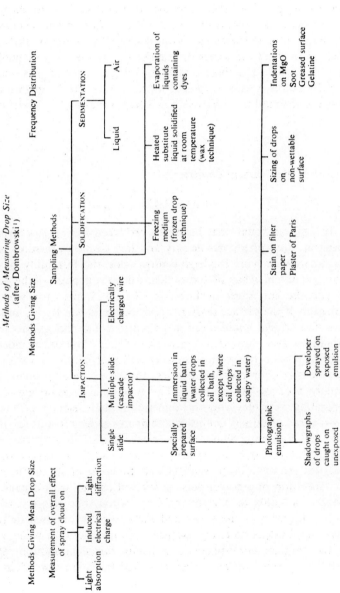

Methods Giving Mean Drop Size

Measurement of overall effect of spray cloud on
- Light absorption
- Induced electrical charge
- Light diffraction

Methods Giving Size

Sampling Methods

IMPACTION
- Single slide
- Multiple slide (cascade impactor)
- Electrically charged wire
- Specially prepared surface
  - Immersion in liquid bath (water drops collected in oil bath, except where oil drops collected in soapy water)
- Photographic emulsion
  - Shadowgraphs of drops caught on unexposed emulsion
  - Developer sprayed on exposed emulsion

SOLIDIFICATION
- Freezing medium (frozen drop technique)
- Heated substitute liquid solidified at room temperature (wax technique)
- Stain on filter paper Plaster of Paris
- Sizing of drops on non-wettable surface
- Indentations on MgO Soot Greased surface Gelatine

SEDIMENTATION
- Liquid
- Air
- Evaporation of liquids containing dyes

Frequency Distribution

Photography of Drops in Flight

Methods of measuring drop size fall into two main groups. The first consists of techniques which result in size frequency distribution of the spray while the second group only permits a mean drop size to be determined. Comprehensive lists of empirical relations for drop size distributions for swirl spray nozzles, blast atomisers and rotary atomisers are given by Dombrowski.[1]

6.4 INTERACTION OF AIR STREAMS AND SPRAYS

Most of the studies on sprays have been for sprays in stagnant air surroundings and only recently has attention been turned to the effect of ambient air velocity on atomisation. Ambient air streams will affect the length and trajectory of the liquid sheet, the degree of atomisation and initial size distribution, but most of all, air streams will affect the trajectories of drops in sprays. The patterns of the air flow through which the liquid spray passes are generally complex and involve recirculation eddies with both forward and reverse flow associated with regions of high shear and high intensity of turbulence. More attention has been directed towards studying the interaction effects and, as a first stage in an extensive series of studies, Mellor et al.[5] have measured droplet trajectories and distributions in a hollow cone spray in a uniform air stream.

In order to obviate the introduction of probes into the spray, high speed photography was used. Drops were photographed in flight as they passed through a 2-mm thick slice of the spray on which the optical system was focused. Two sparks fired at intervals of a few microseconds gave double image photographs of drops from which drop diameters were measured directly from projection of the photograph negatives onto a translucent screen, and drop velocities were calculated by measuring the distance between the two images of the drop and the time interval between sparks. A double flash photograph of the spray sheet emerging from the nozzle is shown in Fig. 6.3. Over a distance of 4 mm the surface of the hollow cone liquid sheet is relatively undisturbed and smooth. The instability in the sheet can be clearly seen as a wave form. Velocities were measured in this region from distances between crests of the two images and the photographs show that the velocity increases with distance from the nozzle. The break up region extends to approximately 10 mm from the nozzle with the formation of ligaments and subsequently drops.

With the exception of only a few of the large drops (200 μm), all drops are spherical beyond the break up region. Instabilities in the liquid sheet and increasing amplitude of waves leads to the formation of drops with a spectrum of sizes, initial velocities and angles. Almost all drops were found to have velocities lower than that of the sheet at break up. The limits of the velocity distributions are very wide, particularly for the

Fig. 6.3. *Disintegration of liquid sheet in pressure jet spray. Double image photograph* *(after Mellor* et al.[5]*).*

smaller drop sizes below 100 μ where velocity varies from that of the sheet at break up (34·5 m/sec) down to the air stream velocity (14·7 m/sec). The distribution appeared to be random for sizes below 100 μ, while for drop sizes above 100 μ velocities are closer to that of the sheet at break up. There is a general tendency for initial drop velocities to decrease as drop size decreases.

Initial angles of drop trajectories were also found to vary above and below that of the liquid sheet (42·2°) and for drop sizes below 100 μ initial angles may vary from 10° to 70°. There is a tendency for angles to approach that of the liquid sheet as drop size increases above 100 μ.

The break up region of the spray is considered as an initial region followed by a transition region where the forces of inertia, drag and gravitation direct the drops towards the developed region of the spray where a balance between these forces allows predictions of trajectories to be made. The concept of a developed region allows extrapolation, from results in the developed region, upstream to an effective point origin from which all droplets may be considered to emanate with the velocity having a single magnitude in direction. The developed region

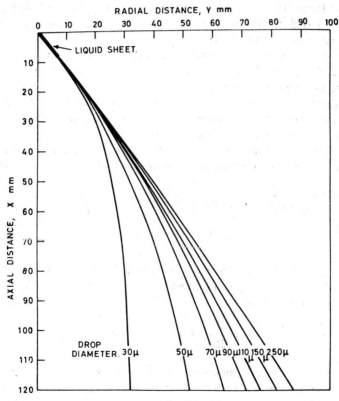

Fig. 6.4. ·*Predicted drop trajectories for pressure jet spray in uniform air stream (after Mellor* et al. [5]).

may be considered to become fully developed when distributions of drop size, velocity and angle of trajectory have 'similar' forms at all subsequent downstream planes perpendicular to the spray axis. This condition will certainly be achieved when all drops attain their terminal velocities, but similarity forms of the distribution curves are expected to be found in regions closer to the nozzle.

Measurements made from photographs taken at a number of positions downstream from the spray show that drag forces play an important role in determining the velocity and angle of drops, particularly in the region after break up. Drops below 30 μ attain their terminal velocity soon after break up while the larger drops having higher inertia do not attain their terminal velocities over the effective working distance of the spray. Predictions of drop trajectories, based on the assumption that all drops emanate from the end of the liquid sheet, are shown in Fig. 6.4. Calculations are based on the equation of motion of isolated particles injected into a uniform air stream. Drag coefficients of drops in sprays can be affected by turbulence in the stream, acceleration, interaction with other drops and internal flow within the drops. No satisfactory experimental data are available which will allow these effects to be taken into account. Consequently, drag coefficients were taken from the standard drag curve for single, solid, smooth spheres moving at constant velocity in a still isothermal fluid which is infinite in extent.

Assuming that evaporation of droplets does not take place, the force balance on a droplet injected into a uniform airstream can be written as

$$\underbrace{\frac{d^3\pi}{6}\rho_p\frac{dV}{dt}}_{} = \underbrace{\frac{d^3\pi}{6}\rho_pg}_{\substack{\text{gravitational}\\\text{force}}} - \underbrace{\frac{d^3\pi}{6}\rho_fg}_{\substack{\text{buoyancy}\\\text{force}}} - \underbrace{C_D\frac{\pi d^2}{4}\rho_f\frac{(\Delta V)^2}{2}}_{\text{drag force}} \qquad (6.5)$$

where C_D is the drag coefficient and ρ_p and ρ_f are droplet density and surrounding fluid density respectively.

For low values of Re,

$$C_D = \frac{1}{\text{Re}}[23 + \sqrt{1 + 16\text{Re}^{4/3}}] \qquad (6.6)$$

This expression of the drag coefficient was used by Mellor[57] in his computer program written for the numerical integration of the equations of motion and for the prediction of droplet trajectories.

Quantitative agreement between predictions and experiments could not be expected in view of the approximations necessary in the predictions, the high degree of randomness in initial conditions of the spray during break up, as well as the necessity to consider statistically average values of velocity and angle obtained by measurements. Figure 6.5 shows the

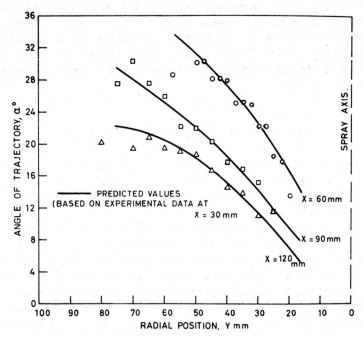

Fig. 6.5. Comparison of predicted and experimental results for angles of trajectory at different radial and axial positions for 80 μm drops (after Mellor et al.[5]).

comparisons of predicted and experimental results for angles of trajectory at different radial and axial positions for 80 μm drops. This figure shows that the air stream causes significant changes in the angle of trajectory of droplets both in the radial and axial directions.

6.5 COMBUSTION OF SPRAYS

The combustion of sprays is a complex process which involves simultaneous heat, mass and momentum transfer and chemical reaction. The chief factors affecting the combustion of spray droplets are (i) drop size, (ii) composition of the fuel, (iii) ambient gas composition, temperature and pressure and (iv) the relative velocity between the droplet and the surrounding gas. Owing to the complexity of the process, it is difficult to obtain detailed and accurate information on the combustion of droplets by means of direct studies on sprays. Because of these difficulties the effects of the above parameters on the burning rates of the droplets have been investigated with single isolated droplets. In addition, a number of experimental investigations have been made to study the effect of interaction between drops during burning. These include experimental studies of stationary, symmetrical droplet arrays, monosize droplet streams and monodisperse sprays.

6.6   SINGLE DROPLET COMBUSTION

A single isolated droplet lends itself well to an experimental investigation on the burning rates as combustion can be carried out under carefully controlled conditions. Furthermore, it is a much simplified problem and is readily amenable to theoretical treatment.

There are two types of droplet combustion—bipropellant and monopropellant. With the former type, which is by far the more important practically, the fuel vapour and oxidant diffuse from opposite directions and a flame forms at the contact surface some distance from the drop. The latter type involves a single reactant system which evaporates and decomposes exothermally, the common examples being hydrazine and ethyl nitrate. The present discussion, however, is confined to bipropellant flames.

THEORETICAL ANALYSIS OF SINGLE DROP COMBUSTION
A general method of predicting the mass burning rate and structure of the flame (both macro and micro) of a burning droplet with a spherically symmetrical flame around it involves the use of the conservation equations developed initially by Hirschfelder and Curtis[6] for one-dimensional, laminar, premixed, gaseous combustion. For a steady state burning droplet with spherical symmetry of combustion, the conservation equations take the following forms:

(i)   Equation of continuity for the total mass flow.

$$\dot{m} = 4\pi r^2 \rho v \qquad (6.7)$$

where $\dot{m}$ = mass rate of flow of fuel vapour leaving the liquid surface, r = radial co-ordinate, v = radial velocity, ρ = density of fuel vapour.

(ii) Equation of continuity for each chemical species.

$$\frac{dY_i}{dr} = \frac{4\pi r^2 \bar{M}_i G_i}{\dot{m}} \tag{6.8}$$

where Y_i = weight fraction of species i in the mass rate of flow, G_i = net chemical reaction rate of i per unit volume, $\bar{M}_i$ = mean molecular weight.

The total net mass flux, $\dot{m} Y_i$, comprises two terms—convection and diffusion. Thus:

$$\dot{m} Y_i = \dot{m} Y'_i - 4\pi r^2 \rho D_i \frac{dY'_i}{dr} \tag{6.9}$$

where Y'_i = weight fraction in convective transport, D_i = diffusion coefficient of i.

(iii) Conservation of energy.

$$4\pi r^2 \bar{k} \frac{dT}{dr} = \dot{m} \left\{ \Sigma \, H_i Y_i - (\Sigma \, (H_i)_{T_l} (Y_i)_{T_l} + Q) \right\} \tag{6.10}$$

where $\bar{k}$ = average value of the thermal conductivity of the mixture, H_i = enthalpy of species i, T_l refers to droplet temperature, Q = latent heat of vaporisation of the drop.

(iv) Conservation of momentum is implied by the usual assumption that combustion takes place under constant pressure.

The following boundary conditions can be used:

$$
\begin{aligned}
T &= T_l & Y_{\text{Fuel}} &= Y_F & \text{at} \quad & r = r_l \\
T &= T_G & Y_{\text{Oxidant}} &= Y_{O_\infty} & \text{at} \quad & r = r_\infty \\
T &= T_F & Y_{\text{Oxidant}} &= 0 & \text{at} \quad & r = r_f \\
& & Y_{\text{Fuel}} &= 0 & &
\end{aligned}
$$

where T_l = droplet surface temperature, Y_F = weight fraction of fuel vapour at the droplet surface, r_l = droplet radius, T_G = temperature of the ambient gas, Y_{O_∞} = weight fraction of oxidant at infinity, r_f = flame radius.

However, because of the boundary conditions imposed and the exponential form of the Arrhenius expression for the chemical reaction rate, analytical solutions of these equations are not possible.

Because of this, in the theoretical analysis an ideal model is considered consisting of a single spherical drop surrounded by a spherical symmetrical flame front as shown in Fig. 6.6a. A steady state transfer of heat and mass

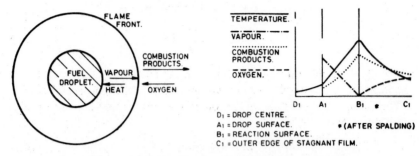

Fig. 6.6(a).   Spherically symmetrical model of a burning droplet.

to and from the droplet is postulated. It is also assumed that chemical reaction rate is infinitely rapid. This model is the basis of a number of theories developed by Godsave,[7] by Spalding[8] and by others.[9-14] The following simplifying assumptions are usually made:

(i) The droplet and flame form concentric spheres. Concentric with the drop and at infinite distance from it lies another outer boundary at which the gas composition is that of the ambient gas. This implies that convection effects are negligible.

(ii) Fuel vapour diffuses from the drop surface to the flame front and oxygen diffuses from the boundary in the ambient gas to the flame surface. The resulting combustion products diffuse from the flame to the surrounding gas.

(iii) Exothermic chemical reaction between the fuel vapour and oxygen takes place at the surface of stoichiometric composition.

(iv) Combustion occurs under isobaric, quasi-steady state conditions.

(v) Chemical reaction occurs instantaneously and hence the reaction zone is infinitely thin. This implies that partial pressures of both fuel vapour and oxygen are zero at the reaction surface.

(vi) Chemical reaction also goes to completion and hence it requires no activation energy.

(vii) A sufficient proportion of the heat produced by combustion is transmitted by conduction to the droplet to provide the latent heat of vaporisation of the fuel while the rest enters the gas stream beyond the stagnant film.

(viii) The droplet temperature is uniform and equal to the boiling point of the liquid.

(ix) Radiation and thermal diffusion effects are negligible.

THEORIES INVOLVING COMBUSTION IN STAGNANT SURROUNDINGS
Some of the more familiar theories based on this model and giving expressions for the mass burning rate of a droplet are now discussed.

STEADY STATE THEORIES
Godsave's[7] was one of the first theoretical attacks on the problem. Using some of the above mentioned assumptions and considering steady state radial heat transfer from the flame to the drop against outward diffusing vapour, he was able to solve the resulting Fourier–Poisson equations to give the following expression for the mass rate of evaporation. This, of course, is equal to the mass burning rate of the droplet. Thus:

$$\frac{dm}{dt} = \frac{2\pi k d \ln\{1 + [c(T_f - T_s)/Q]\}}{c[1 - (d/d_f)]} \quad (6.11)$$

where dm/dt = mass burning rate or mass rate of evaporation, k and c are the mean values of thermal conductivity and specific heat between the drop surface and the flame front, T_f = flame temperature (K), T_s = surface temperature of the drop, d = drop diameter, d_f = flame diameter, Q = latent heat of vaporisation at the drop surface temperature. The radiation term in eqn. (6.11) has been neglected.

However, as no use was made of the assumptions of oxygen diffusion to the flame and location of the reaction surface, the position of the flame surface relative to the drop and flame temperature are not given by this theory. Experimental values of d_f and calculated values of T_f have to be used to calculate dm/dt from eqn. (6.11).

Long[9] extended this theory by considering a heat balance between the flame and the droplet surface and an oxygen mass balance beyond the flame. His final expression of the mass burning rate comprises a heat and mass transfer term. This expression does not contain the term d_f and is given by

$$\frac{dm}{dt} = \frac{2k}{cd} \ln\left\{1 + \frac{c(T_f - T_s)}{Q} + \frac{\rho_O D_O}{16\chi d} \ln\left(\frac{1 + N_\infty}{\Omega}\right)\right\} \quad (6.11a)$$

(however, for typical fuels, the first two terms in the bracket are large compared with the third term and hence it can be concluded that the combustion is controlled by heat transfer)

where ρ_O = mean density of oxygen beyond the flame,
$\quad D_O$ = diffusion coefficient of oxygen beyond the flame,
$\quad N_\infty$ = mole fraction of oxygen in the bulk surrounding gas,

χ = number of moles of combustion products derived from stoichiometric combustion of unit mass of fuel with oxygen,

Ω = number of moles of oxygen required for stoichiometric combustion of unit mass of fuel.

Goldsmith and Penner's theory[10] is basically a refinement and generalisation of Godsave's. Use was made of all the assumptions of the spherico-symmetrical model. Equations of conservation of energy and of continuity were applied to spherical shells of diameters d'_1 and d'_2, where $d < d'_1 < d_f$ and $d_f < d'_2 < d_\infty$ (d = drop diameter, d_f = flame diameter, d_∞ = diameter of a spherical shell at infinite distance).

These equations were solved analytically to give the mass burning rate, dm/dt, as an eigenvalue. In addition, the expressions for T_f, d_f and Y_F (where Y_F = weight fraction of the fuel vapour at the droplet surface) were obtained. In these derivations, k and c were assumed to be linear functions of temperature

$$k = k_i \left(\frac{T}{T_s}\right) \tag{6.12}$$

$$c = a' + b'T \tag{6.13}$$

The mass burning rate,

$$\frac{dm}{dt} = \frac{2\pi k_i d}{T_s b'[1 - (d_f/d)]} \left\{ \ln\left[1 + \frac{(T_f - T_s)}{Q}\left(\frac{a' + b'}{2}\right)(T_f + T_s)\right] \right.$$
$$\left. - \frac{a'}{\varepsilon} \ln\left[\frac{(a' + b'T_f - \varepsilon)(a' + b'T_s + \varepsilon)}{(a' + b'T_f + \varepsilon)(a' + b'T_s - \varepsilon)}\right] \right\} \tag{6.11b}$$

where

$$\varepsilon = \left[a'^2 - 2b'\left(Q - \frac{b'}{2}T_s^2 - a'T_s\right)\right]^{1/2}$$

The theory also predicts that d_f/d should be constant for given values of the physico-chemical parameters.

The same basic equations have been used by others to obtain expressions for the mass burning rates. Usually assumptions have been made about the properties of the vapour–gas mixture between the droplet surface and the flame. For example, Wise et al.[11] assumed that ρD and c/k were constants while Hottel et al.[12] used a log mean value of k/c in this region.

Expressions have also been derived by some of these workers for flame temperature, the mass fraction of fuel vapour at the droplet surface and the ratio of flame to drop diameter. The following simple expressions obtained by Wise *et al.*[11] are typical of the form of the relations:

$$T_f = T_s + \frac{H - Q}{c} - \left\{ \frac{[(H - Q)/c] + (T_s - T_G)}{[1 + (Y_{O_\infty}/i)]} \right\} \qquad (6.14)$$

$$\frac{d_f}{d} = \frac{\ln [(H/Q)(Y_{O_\infty}/i)]}{\ln [1 + (Y_{O_\infty}/i)]} \qquad (6.15)$$

$$Y_F = 1 - \frac{Q}{H} \left(1 + \frac{i}{Y_{O_\infty}} \right) \qquad (6.16)$$

where H = heat of combustion, T_G = temperature of the ambient gas and Y_{O_∞} = weight fraction of oxidant at infinity.

Further, by this method it is also possible to deduce the temperature and composition profiles in the flame surrounding the drop. This has been achieved by Goldsmith and Penner[10] and Wise *et al.*[11] Typical profiles are shown in Fig. 6.6b.

Spalding[8] approached the problem in a slightly different manner. He postulated that the combustion takes place in a stagnant film adhering to the surface of the drop, even when convection is present. He further postulated an analogy between heat and mass transfer processes and showed that the thickness of the stagnant film for combustion can be

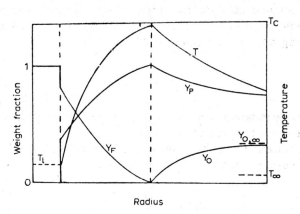

Fig. 6.6(b). Theoretical temperature and composition profiles of a droplet burning in absence of convection (after Williams[36]).

predicted from heat transfer coefficients. This knowledge permits formulation and solution of the following equations governing various processes occurring within the film:

(i)   diffusion of oxygen from the outer edge of the film to the flame surface;

(ii) conduction of heat from the flame to the drop;

(iii) heat transfer from the flame to the main gas stream.

From these, Spalding obtained an expression of the mass burning rate, T_f and d_f. However, when combustion takes place in stagnant surroundings, the film thickness is infinite. For these conditions, he obtained the following expression of the mass rate of evaporation which is equal to the mass burning rate of the droplet:

$$\frac{dm}{dt} = 2\pi d\rho D \ln (1 + B) \qquad (6.11c)$$

where B is termed the transfer number and is defined by

$$B = \frac{H}{Q} \frac{M_{O_2}}{i} + \frac{c(T_G - T_s)}{Q} \qquad (6.17)$$

The quantity ρD can be replaced by k/c if Le is assumed to be unity. Hence we have

$$\frac{dm}{dt} = 2\pi d \frac{k}{c} \ln (1 + B) \qquad (6.11d)$$

One important conclusion that can be drawn from the above theories is the following: $dm/dt \propto d$, since all other terms in the expressions of the mass burning rate can be assumed to be constants.

Hence,

$$\frac{dm}{dt} = K_1 d \qquad (6.11e)$$

where K_1 is a constant which is a function of the physical properties of the fuel only and d is the drop diameter.

Further, since the mass burning rate equals the time rate of change of mass of the droplet, eqn. (6.11e) can be written as

$$\frac{d}{dt} \left(\frac{\pi}{6} d^3 \rho_l \right) = K_1 d \qquad (6.11f)$$

where ρ_l = density of the liquid, which is assumed constant.

Differentiating and rearranging, we have,

$$\frac{\pi}{2} \rho_l d \, \mathrm{d}(d) = K_1 \, \mathrm{d}t \qquad (6.11\mathrm{g})$$

Integration of eqn. (6.11g) for the boundary conditions d_0 at $t = 0$ and d at time t, leads to

$$d_0{}^2 - d^2 = -\frac{4K_1}{\pi \rho_l} t \qquad (6.18)$$

or

$$d_0{}^2 - d^2 = \lambda t$$

where

$$\lambda = -\left(\frac{4K_1}{\pi \rho_l}\right) \qquad (6.19)$$

From eqn. (6.19), it is evident that λ is constant for a particular fuel. It has usually been termed the evaporation or burning constant.

From eqns. (6.11e) and (6.19), it can be shown that λ is related to $\mathrm{d}m/\mathrm{d}t$ by the following equation:

$$\lambda = \frac{-4(\mathrm{d}m/\mathrm{d}t)}{\pi \rho_l d} \qquad (6.19\mathrm{a})$$

By using eqn. (6.19a), the expressions for λ can be obtained from the quasi-steady state theories, *i.e.* from eqns. (6.11), (6.11a), (6.11b) and (6.11d).

Thus, for example, from eqn. (6.11d) we get the following expression for λ:

$$\lambda = -\frac{8k \ln (1 + B)}{\rho_l c} \qquad (6.19\mathrm{b})$$

In all the theories outlined so far, factors such as chemical kinetics and all the transient processes (ignition, heating up, etc.) have been neglected. This allowed the analytical solutions of the transport and conservation equations to be obtained. However, some attempts have been made to include these complex processes in a generalised theory of droplet combustion.

UNSTEADY STATE THEORIES

Kotake and Okazaki[15] replaced the quasi-steady state model with an unsteady state, diffusion controlled model, but retained most of the assumptions of the older theory.

The equations of mass concentration, velocity and temperature of the gas surrounding the droplet were written in finite difference form with respect to time and space. These were then solved numerically using stepwise increments of time. Their main conclusions are as follows:

(i) the values of $d(d^2)/dt$ change in the initial period of combustion and then approach constant values which are equal to those obtained from quasi-steady state models;

(ii) the calculated values of T_f and d_f/d are much smaller than those of the quasi-steady state model;

(iii) the values of d_f/d and T_f change during combustion.

The phenomena of ignition and of penetration of thermal waves into the liquid following ignition are transient, hence quasi-steady state theories cannot explain them. A number of theoretical analyses have been made of these processes.

Peskin *et al.*[16-18] have developed a number of theories of ignition based on finite chemical kinetics and on the concept of a 'modified flame surface' as defined by the 'Dirac delta function'. All these theories postulate that ignition is a process involving transition from chemically controlled to diffusion controlled combustion. Theoretical work on the phenomenon of 'heating up' shows that an appreciable period of time elapses before the centre of the drop attains the surface temperature of the drop. Wise and Ablow,[19] solving the equation of heat conduction numerically, derived temperature distributions within the droplet as a function of the group $(\lambda/8\alpha)$, λ being the burning constant and α the thermal diffusivity of the liquid.

When $\lambda/8\alpha$ is 1, the centre of the droplet attains 90% of its surface temperature during the first half life of the droplet but only 12% if $\lambda/8\alpha$ is 2.

THEORIES INVOLVING FINITE CHEMICAL KINETICS

In order to test the accuracy of the assumption of an infinite chemical reaction rate and to examine the conditions under which chemical kinetics might affect the combustion rate, several attempts have been made to solve the conservation equations without this assumption.

Lorell *et al.*[20] considered a finite reaction rate of the bimolecular Arrhenius form; rate $= A'$ (fuel) (oxygen) exp $(-E/RT)$, and solved the conservation equations numerically for an ethanol drop of diameter 0·4 mm burning in air and in oxygen. Values of E of 0, 4·815, and 9·63 kJ/kg were assumed. The results did not differ greatly from those assuming an infinite chemical reaction rate. They have concluded that chemical kinetics can only seriously affect the combustion rate when either the droplet size or the ambient pressure is very low.

Williams[21] has used the Shvab–Zeldovich procedure to obtain the expression

$$\frac{dm}{dt} = 2\pi d\frac{k}{c}\ln(1+B) - \frac{2\pi dkHY_{O_i}}{cQ} \qquad (6.11h)$$

where

$$B = \frac{H}{Q}Y_{O_i} + \frac{c}{Q}(T_G - T_s) \qquad (6.17a)$$

in which the second term is a correction term for finite chemical kinetics.

THEORIES INVOLVING COMBUSTION IN CONVECTIVE GAS FLOW
All the theories presented above have ignored the effects of convection. In practice, however, the droplets always experience convection, either natural, forced or both. Theoretical attempts to estimate the influence of convection on combustion will now be reviewed.

Natural convection effects
The presence of natural convection in droplet combustion destroys the spherical symmetry of combustion and distorts the flame from its spherical shape (Fig. 6.6c).

Calculation of the mass burning rate of a droplet experiencing natural convection would require solutions of the equations of conservation of energy, of continuity and of the Navier–Stokes equations.

The boundary conditions for the latter would be complex, and non-linearity would arise owing to the exponential form of the Arrhenius

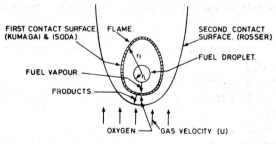

Fig. 6.6(c). Ideal model of a burning droplet in presence of convection.

equation and the inertia terms of the Navier–Stokes equations. Under these conditions an analytical solution is not possible.

Due to this problem, the effect of natural convection on the burning rate has been taken into account by means of empirical correlations. Two of these are noted below:

(i) Spalding[8]

$$\dot{M}_N = 0{\cdot}45 \frac{k}{c} B^{3/4} \left(\frac{gd^3}{\alpha_1{}^2}\right)^{1/4} \tag{6.20}$$

which is valid for $0{\cdot}25 < B < 3$. (k, α_1 were evaluated at room temperature.)

(ii) Agoston et al.[22]

$$\dot{M}_N = \dot{M}_s \left(1 + \frac{0{\cdot}20}{B^{0{\cdot}44}} \, Gr^{0{\cdot}3}\right) \tag{6.20a}$$

A log mean temperature was used for calculating the physical properties and Gr.

Several attempts have also been made to evolve theories which will accurately predict the flame shape of a droplet burning under free convection conditions. A common assumption is that a burning droplet behaves as a point source of fuel vapour which flows radially into a uniformly moving gas stream. Isoda and Kumagai[23] postulated that a contact surface exists between the air and vapour flows. On calculating the position of this surface, they found that it coincided with the lower part of the flame front (Fig. 6.6c). Rosser,[24] on the other hand, defined a contact surface as the streamlines which separate the oxidant from the combustion gases. Mass transfer was assumed to occur by diffusion across the contact surface (Fig. 6.6c). Schlieren photography was used to determine the position of this surface. This was then related to the position of the flame.

Forced convection effects
Forced convection can affect the combustion of droplets in a number of ways:[25]

(i)   there is a considerable deviation of the flame from spherical symmetry;
(ii) the film thickness is not infinitely large;
(iii) the local heat transfer and evaporation rates change over the drop surface;
(iv) the mass burning rate is higher than that predicted by the spherico-symmetrical model.

The variation of the heat transfer rate over a droplet experiencing forced convection is shown in Fig. 6.7.

For the sake of simplicity, however, most of the correlations of the burning rates neglect these changes and use an average value for the whole droplet. Some of the empirical and semi-empirical correlations of the mass burning rates of the droplets under forced convection conditions are presented in Table 6.3.

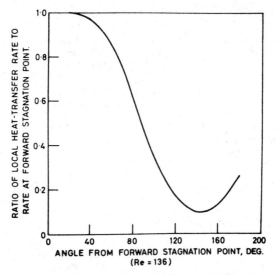

Fig. 6.7. The variation of local heat transfer rate over a drop surface under forced convection conditions (after Graves and Bahr[25]).

The increase in burning rate with convection shown in Table 6.3 is due to the reduction in the thickness of the stagnant film beyond the flame. This increases the oxygen concentration at the flame surface and so decreases the distance between the flame and the drop surface.

A limited number of theoretical analyses on flame extinction have been made.

Spalding[30] has proposed that extinction of a droplet will occur when the mass rate of evaporation exceeds the maximum possible rate of consumption of the vapour in the flame. He regards the maximum rate of combustion as being that of a premixed stoichiometric flame.

Peskin *et al.*,[16-18] using the 'linearised flame surface model', concluded that the flame extinction is a transition from diffusion controlled combustion to one of chemical control.

TABLE 6.3

Various Correlations of the Mass Burning Rate or Mass Rate of Evaporation Under Forced Convection Conditions

Name of investigators	Experimental conditions	Empirical or semi-empirical correlations	Comments
Frossling[26]	At room temp., nitrobenzene, naphthalene, water $2 < \text{Re} < 800$	$\dot{M}_F = \dot{M}_s(1 + 0.276\,\text{Re}^{1/2}\,\text{Sc}^{1/3})$	
Ranz and Marshall[27]	Up to 473K, ambient temperature	$\dot{M}_F = \dot{M}_s(1 + 0.3\,\text{Re}^{1/2}\,\text{Sc}^{1/3})$ For high temperature evaporation or droplet combustion it is customary to replace Sc by Pr so that M^F is given by $\dot{M}_F = \dot{M}_s(1 + 0.3\,\text{Re}^{1/2}\,\text{Pr}^{1/3})$	
Agoston et al.[22]	At 2200 to 2920K, ethanol and methanol burning from a model sphere	$\dot{M}_F = \dot{M}_s(1 + 0.24\,\text{Re}^{1/2})$	Pr was assumed constant and equal to 1
Spalding[8]	Model sphere technique, $800 < \text{Re} < 4000$, $0.6 < B < 5$	$\dot{M}_F = 0.53B^{3/5}\,\text{Re}^{1/2}\,\dfrac{\mu}{d}$	B is the transfer number and is defined by eqn. (6.17)
Eisenklam et al.[28]	Freely moving burning drops, $0.1 < \text{Re} < 5$	$\overline{\text{Nu}} = \dfrac{4.9}{(1 + B)}\,\text{Re}^{0.4}$	
How[29]	For burning or evaporating drops	$\dot{M}_F = \dot{M}_s(1 + 0.22\,\text{Re}^{1/2})$	

Polymeropoulos and Peskin[18] extended this analysis and calculated extinction behaviour of the droplets over a wide range of conditions. According to their results, a plot of mass burning rate vs decreasing temperature shows a point on the curve with an infinite slope. Upon reaching this temperature there is a discontinuity in the decrease of burning rate. The authors postulate that this point corresponds to the extinction of the flame.

Another aerodynamic effect, of considerable importance with drops under forced convection conditions, is the drag coefficient of both burning and unignited drops. The fundamental studies of this phenomenon have been carried out with single inert spheres and non-evaporating droplets. The more important results are discussed in reference 31. The important theories and correlations of C_D (drag coefficient) for evaporating and burning drops may be summarised thus.

The equation of motion of an evaporating drop differs from that of an inert sphere by a term which accounts for the inertial force contributed by the mass flux from the surface of the drop. This affects the drag force on the drop.

Spalding[32] showed that momentum of the effusing vapour alters the velocity profiles within the boundary layer and so leads to the increase in the thickness of this layer. As a result, shear force on the surface of the drop (*i.e.* frictional drag) is reduced.

In the case of a burning drop, due to expansion in the flame, combustion gases fill in the low pressure regions within the wake and thus reduce the form drag[33] (*i.e.* the integral of pressure distribution over the surface of the drop).

Analytical solutions of the Navier–Stokes equations for drops undergoing mass transfer have only been obtained for a number of highly idealised systems at $Re \leq 1$.

Muggia[34] determined theoretically the flow around an evaporating sphere at $Re \leq 1$ by coupling Oseen's linearised form of the Navier–Stokes equations with the diffusion equation. Using the assumptions of uniform surroundings, uniform drop temperatures, zero tangential velocity and finite radial velocity at the drop surface, he obtained

$$\overline{C_D} = \frac{24}{Re} \left[\frac{2 + B}{(2 + \frac{3}{4}B) - \frac{3}{16} Re (2 + B)} - \frac{B (4 + Re)}{3 (2 + B)} \right] \quad (6.21)$$

where

$$B = c \frac{(T_G - T_s)}{Q}$$

T_G = temperature of the ambient gas, T_s = drop surface temperature, Q = latent heat of vaporisation, c = specific heat of gas at constant pressure.

The values of $\overline{C_D}$ predicted by eqn. (6.21) are considerably lower than those of the standard curve.

Crowe et al.[35] studied analytically the effects of burning on the drag coefficients of accelerating particles at $250 < Re < 1600$ and found a decrease in drag when the ratio of the mass flux from the surface to the mass flux in the free stream was greater than 0·025.

Spalding[32] studied experimentally the effect of intense mass transfer on the drag coefficients of flat plates in laminar flow. He found that his results of drag coefficients could best be correlated by the stagnant film theory, *i.e.* the following relation is valid:

$$\frac{\overline{C_D}}{C_D} = \frac{\ln (1 + B)}{B} \tag{6.21a}$$

where $\overline{C_D}$ = drag coefficient in presence of mass transfer and C_D = standard drag coefficient.

Eisenklam *et al.*[28] investigated the effect of intense mass transfer and flame on the drag coefficients of liquid drops and correlated their results of evaporating and burning drops by the following boundary layer theory:

$$\frac{\overline{C_D}}{C_D} = \frac{1}{1 + B} \tag{6.21b}$$

where the range of Re was $0.1 < \text{Re} < 3$ for burning drops and $0.1 < \text{Re} < 40$ for evaporating drops.

Both eqns. (6.21a) and (6.21b) give values of $\overline{C_D}$ considerably lower than the standard curve.

EFFECT OF TURBULENCE INTENSITY

No systematic theoretical analyses of the effects of turbulence on the burning rates of the drops have been carried out. However, How[29] has pointed out that in turbulent systems with rms values of the fluctuating velocity around 20% of the main stream velocity, turbulence increases the burning rates of the drops by improving: (a) the mixing between adjacent fuel-rich and fuel-deficient gas pockets; and (b) the heat transfer between the zones at different temperatures. For systems with higher levels of turbulence, however, some 'scrubbing' might occur which would lower the vapour concentration around the droplets. All these effects tend to increase the burning rate.

The predictions of the various theories are listed in the following section so that they can be compared with the experimental results.

SUMMARY OF THEORETICAL PREDICTIONS

Quasi-steady state theories
(i) The mass burning rate of a droplet is proportional to the first power of the droplet diameter. This implies that the square of the diameter decreases linearly with time.

(ii) The ratio of flame diameter/drop diameter is constant during combustion.

(iii) The flame temperature is constant and equal to the adiabatic flame temperature. The flame is formed at the locus where the ratio of fuel to oxidant flows is stoichiometric.

(iv) The weight fraction of fuel vapour at the drop surface is less than unity, hence the surface temperature is less than the boiling temperature of the fuel.

(v) The mass burning rate is insensitive to pressure.

(vi) The physical properties which have a dominant effect on the combustion rate are: the latent heat of vaporisation of the fuel, the specific heat and thermal conductivity of the vapour/gas mixture and the heat of combustion.

(vii) Chemical kinetics have no effect on the combustion rate provided the drop size and the ambient pressure are large.

Non-steady state theories

(viii) The values of the slopes of d^2 vs t curves change in the initial period of combustion and then attain constant values equal to those of the quasi-steady state theories.

(ix) The values of droplet surface temperature, of flame to drop diameter ratios and of flame temperatures change during combustion.

(x) A finite period of time must elapse before the centre of the droplet attains the surface temperature of the droplet.

(xi) The ignition process is a transition from chemical to diffusion controlled combustion. Extinction is the reverse of this process.

EXPERIMENTAL WORK ON SINGLE DROPLETS

Single droplet combustion has been investigated using:

(i) suspended drops;
(ii) model spheres;
(iii) drops in free flight.

An extensive amount of data on the effects of drop size, fuel characteristics, ambient temperature, pressure and gas composition, etc. on the burning rates has been derived from these investigations. The more important results are reviewed by Williams[36] and by Nuruzzaman.[37] A limited number of experimental studies of the effect of forced convection on the burning rate have also been made.

TABLE 6.4

The Effect of Ambient Gas Velocity on the Burning Constant

Experimental technique	Relative gas velocity	Change in the burning constant	Extinction velocity
Suspended drop technique (Goldsmith[38])	Ambient gas velocity was increased from 0 to 400 mm/sec	36% increase in λ	Flame extinction occurred at a velocity greater than 400 mm/sec
Suspended drop technique (Kumagai[39])	Ambient gas velocity was increased to 450 mm/sec	38% increase in λ	Flame extinction occurred at higher velocity (450 mm/sec)
Suspended drop technique[40] (furnace temp. = 973K Masdin)	Ambient gas velocity was increased from 0 150 mm/sec	10% increase in λ	Auto ignition failed to occur at higher velocity

As Table 6.4 shows, the burning constants are increased by the increase in relative velocity between a droplet and the ambient gas.

Empirical correlations have also been obtained between mass burning rate and Re (see Table 6.3).

Kumagai and Isoda[41] have obtained the following correlation for single droplets burning in vibrating air fields:

$$\lambda_z = \lambda_s + K'_1 f' p_1{}^2 (K'_2 - f' p_1{}^2)$$

where λ_s = burning constant of the drop in stagnant medium, λ_z = burning constant with vibrations, K'_1, K'_2 = experimental constants, f' = frequency of vibration and p_1 = amplitude of air vibration.

The burning constant, λ_z, can be increased by up to 15% over the still air value, apparently by an increase in the gas diffusivity. Further increase in the vibration amplitude reduces the burning rate. This is attributed to the deformation of the flame boundary.

Forced convection can radically alter the type of flame supported by a droplet.

Spalding[30] observed envelope flames at low stream velocities. At speeds in excess of the extinction velocity, he reported that a flame could be stabilised behind the drop, which he termed a wake flame. An envelope flame is a diffusion flame, while a wake flame is of a premixed type. Udelson[42] has noted a third type, namely a flame which is stabilised within the boundary layer at the side of a drop.

Eisenklam and Arunachalam[33] have found that the drag coefficients of burning drops are strongly influenced by the type of flame associated with the drops. They found that the values of C_D for drops with side flames were less than the standard values, whereas those of the drops supporting envelope flames were higher. These values were calculated on the basis of approach properties, *i.e.* the properties of the undisturbed atmosphere in the chamber.

THE COMBUSTION OF RESIDUAL FUELS

With residual fuels, solid residues are formed after the combustion of the volatile constituents.[12,40,43,44]

Hottel *et al.* presented their data for burning residual fuel oil droplets in the form of a plot of $dm^{2/3}/dt$ vs t. The slope decreases abruptly after 90% of the droplet mass has been consumed. This point is believed to coincide with the completion of the volatile combustion and the onset of residue burning.

Masdin and Thring[40] reported that the cenospheres formed during combustion burn at between one-third and one-tenth of the rate of the volatile portion of the fuel.

They suggested the relation

$$d_c = 0.66\, d_0{}^2 \tag{6.22}$$

for the combustion of pitch creosote droplets in air at 973K.

6.7   THE COMBUSTION OF DROPLET ARRAYS, STREAMS AND MONODISPERSE SPRAYS

A number of experimental investigations have been made to study the effect of interaction between drops during burning.[45-48]

These include experimental studies of two, five and nine suspended drops with varying initial diameters and initial separations. The results for the two-droplet system[45] show that d^2 vs t is a straight line for both the droplets during combustion. It is also found that as the initial separation of two drops is reduced from a high value where interference effects are negligible, the absolute value of λ (where $\lambda = d_0{}^2/t_b$) increases to a maximum and then diminishes. As the drops are brought closer together, heat losses are reduced with a corresponding increase in λ, but, as the separation is further reduced, the amount of oxygen available is diminished which in turn reduces the value of the burning constant. Work on symmetrical five-[45] and nine-droplet arrays[46] shows that λ is higher for the centre drop than for two drops close together. These results suggest that the absolute value of λ is dependent on array geometry.

Experiments have also been reported on the rate of flame propagation, through one-dimensional arrays, of octane droplets under both natural[47] and forced[48] convection. The following correlations have been obtained:
  (1) Reichenbach *et al.*[47]

$$\tau = K_2 I_{D_0}{}^{-3/2} \tag{6.23}$$

for droplets burning under natural convection (τ = flame propagation time, K_2 = a constant and I_{D_0} = the immersion depth which is defined as the minimum depth to which an unburned droplet is immersed in the hot gas region of the adjacent burning droplet).
  (2) Iinuma[48]

$$\tau = \frac{K_3}{(I_{D_0} + K_4 V - S)^{3/2}} \tag{6.23a}$$

Fig. 6.8. Combustion of a monosized droplet stream (after Nuruzzaman et al.[49]).

for droplets under forced convection (K_3, K_4 are constants, I_{D_0} = the immersion depth under natural convection as defined above, V = air velocity and S = droplet spacing).

In order to determine the combustion rates of drops of very small sizes, typical of those found in industrial practice, and to investigate the effect of droplet interaction, Nuruzzaman et al.[49] studied the combustion of freely suspended, monosized droplet streams. In these studies, streams of monosized, uniformly spaced droplets were produced by a hypodermic needle which was made to vibrate at resonance frequency by the electromagnetic vibration of the diaphragm of an earphone. The size of the drops could be varied within the range 10 to 500 µm and the technique also enabled the variation of inter-drop spacing within a wide range. When ignited, these streams produced stationary, self-supporting flames. The photographs of the moving drops in the stream were taken with a 1–2 µsec

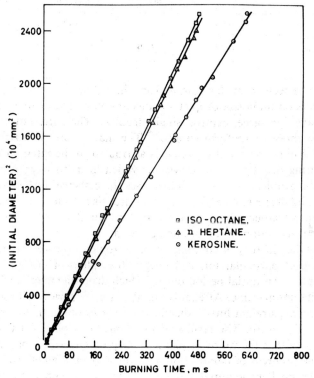

Fig. 6.9. Burning time of a droplet as a function of the square of initial diameter and of fuel type in monosized droplet streams (after Nuruzzaman et al.[49]).

duration flash while the camera shutter was left open for 1/25 sec for recording the flame. The change in droplet diameter during combustion was determined from photographs, as shown in Fig. 6.8. The corresponding time base was obtained from the frequency of vibration of the needle.

In Fig. 6.9, the squares of initial diameters have been plotted against corresponding burning times for monosized droplet flames of kerosine, iso-octane and n-heptane. The slopes of these lines are the burning constants [λ in eqn. (6.18)]. The average values of these constants along with those obtained from single droplets, droplet arrays and monodisperse sprays are given in Table 6.5.

TABLE 6.5

Burning Constants λ (mm^2 sec^{-1})

Fuel	Single drops[7]	Two-drop array[45]	Monosize stream[49]	Uniform spray[50]
Kerosine	0·96	—	0·405	—
Iso-octane	0·95	—	0·548	—
n-Heptane	0·97	0·88	0·529	0·47

As can be seen from this comparison, the values determined from the experiments on monosize droplet streams are only 50% of those reported by Godsave[7] on large, captive, single droplets. They are, however, very similar to those of uniform sprays.[50] The main cause of observed low values of λ in streams and in sprays is attributed to the interaction of the neighbouring droplets in these systems which leads to oxygen depletion around the droplets. This in turn increases the distance of the flame from the drop and decreases the flame temperature. As a result of these changes, temperature gradients between the drop and the flame are reduced with consequent reduction in the burning constant.

The detailed history of combustion of a droplet, however, showed that, for a drop of particular initial diameter, the values of $d(d^2)/dt$ increase appreciably in an initial period during which drop diameter is reduced to 90% of its initial value. After this initial unsteady period, all the d^2 vs t curves become straight lines indicating that the slopes in these regions are constant (Fig. 6.10). The results of the flame sizes obtained from these studies show that the ratios of flame to drop diameter increase markedly during the progress of combustion and, for a given diameter in different flames, the smaller the initial diameter of the drop the greater is the value of d_f/d.

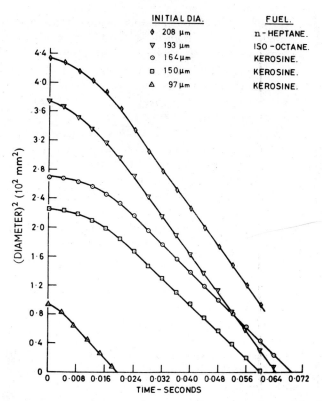

INITIAL DIA.		FUEL.
◊	208 μm	n - HEPTANE.
▽	193 μm	ISO - OCTANE.
◦	164 μm	KEROSINE.
□	150 μm	KEROSINE.
△	97 μm	KEROSINE.

Fig. 6.10. *Variation of drop diameters during combustion in monosized droplet streams (after Nuruzzaman* et al.[49]).

These results contradict the quasi-steady state theories according to which the ratio d_f/d is constant and independent of drop diameter. These are, however, explained by Kotake and Okazaki's[15] model of droplet combustion, according to which a temperature gradient exists within the drop for an appreciable part of the lifetime of the drop and also the combustion is an unsteady diffusion-controlled phenomenon. This theory predicts an increase in the ratio of d_f/d during combustion. This is therefore in line with the observed phenomena. These workers suggested that one possible explanation for constant values of $d(d^2)/dt$, even when the flame to drop diameter changes markedly, is that the changes in this ratio in these cases counterbalance those in flame temperatures thus maintaining a constant temperature gradient between the flame front and the drop surface.

Values of Nusselt number for burning drops determined by these workers[49] were 80–90% lower than those for drops with negligible mass transfer. This large decrease was explained by the fact that in a burning drop the heat transfer to the drop occurs against outward evaporative flow of vapour. Because of this, the temperature distribution within the boundary layer is modified and this changes the viscosity of the gas in this region. Further, the momentum of the effusing vapour alters the velocity profiles within this layer. The net effect of these changes is that the thickness of the boundary layer surrounding the drop is increased.

In order to determine the relationship between the 'evaporation constants' determined from single droplet combustion and those from sprays, a number of workers studied experimentally the evaporation rates of uniform sprays.

Bolt and Boyle[50] studied photographically the burn away of droplets in a nearly uniform spray. They concluded that there was no single burning rate for all the droplets, and that the measured burning rates varied widely owing to the different ambient conditions experienced by the droplets. However, the plot of mean diameter of the drops at successive zones of the flames against time gave straight lines. The values of the slopes of these lines were 50% of those reported by Godsave on large, single, captive drops.

Using tetralin aerosols of monosized drops, Burgoyne and Cohen[51] investigated the effect of drop size on the combustion characteristics. They found that with droplets $< 10 \ \mu m$ in diameter, an aerosol–air mixture burns like a premixed gas. With droplets $> 40 \ \mu m$ in diameter, however, the droplets burn in their own enveloped flames and obey the d^2 law [eqn. (6.18)]. Between 10 and 40 μm, the combustion is transitional. The average values of λ [where $\lambda = \mathrm{d}(d^2)/\mathrm{d}t$] obtained by Burgoyne and Cohen are about half of those of Godsave.

6.8   TWIN FLUID ATOMISED LIQUID FUEL DIFFUSION FLAMES

One of the earliest results of the work at Ijmuiden[52] showed that, for a given fuel input, the factor that most influences flame characteristics such as flame length, temperature distribution or radiation from a fluid atomised turbulent diffusion flame is the burner reaction thrust of the fuel jet (linear momentum flux of fuel plus atomising agent).

In flames with higher values of the jet momentum flux, mixing is quicker between fuel and combustion air entrained by the jet, which in turn leads to higher rates of heat release. The flame contains less soot and has a lower emissivity.

As a result of a systematic series of trials carried out at Ijmuiden, in which concentration and temperature distributions along the flame and radiation from the flame were determined as a function of input variables such as type of atomising agent, atomising agent mass and jet momentum, it was shown that the appropriate objective of burner design optimisation

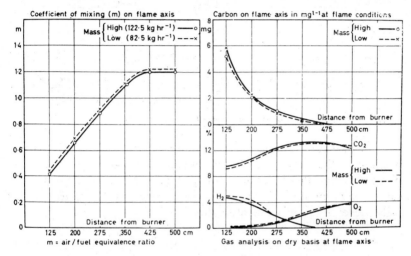

Fig. 6.11.   *Effect of atomising agent mass flow rate.*[52]

is the production of jet momentum flux at the lowest expense in atomising agent mass. Different types of twin fluid (blast) atomising burners can therefore be best compared for the two criteria (a) total jet momentum flux produced, and (b) atomising agent mass required. Figures 6.11 to 6.13 illustrate these points. They show the coefficient of mixing m, the amount of solid carbon and the concentrations of CO_2, O_2 and H_2 along the flame for the effect of the input variables. In the mixing coefficient, the local air/fuel mixture ratio calculated from the composition of a sample is related to the stoichiometric ratio.

Figure 6.11 shows that [52] the atomising agent mass has no effect on the above flame parameters if the jet momentum flux is maintained constant.

Figure 6.12 represents the effect of the type of atomising agent. Mixing is somewhat faster with air as atomising agent and this in turn is shown to

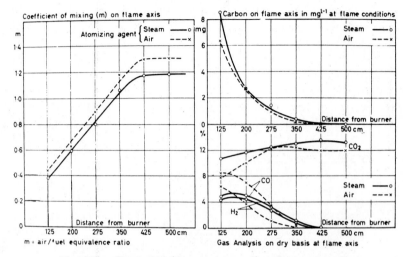

Fig. 6.12. *Comparison between steam and air for atomising.*[52]

have some influence upon the decay of soot concentration and gas composition in the flame close to the burner. The significant effect of the burner reaction thrust is illustrated by Fig. 6.13. The decay of the solid carbon concentration along the flame is considerably faster with the higher level of the jet momentum flux.

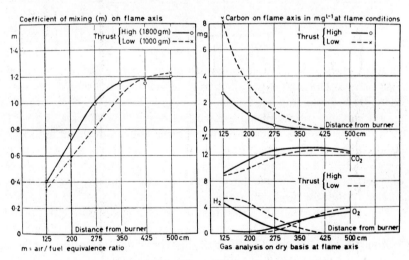

Fig. 6.13. *Effect of burner thrust.*[52]

Figures 6.14(a) and (b) show the result of the optimisation of burner design following the above experiments. The simple convergent nozzle in Fig. 6.14(a) was replaced by a convergent–divergent nozzle designed by BISRA which enabled a saving of approx. 20% atomising steam while maintaining the same jet momentum.

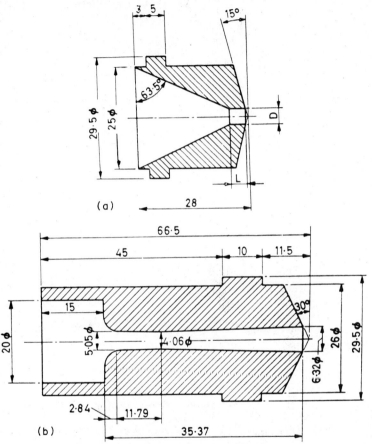

Fig. 6.14. (a) Simple convergent nozzle. (b) Convergent–divergent nozzle.[52] Dimensions in millimetres.

6.9 PRESSURE JET OIL FLAMES

The International Flame Research Foundation carried out a systematic study of pressure jet oil flames in the experimental furnace at Ijmuiden

(Holland), during the years 1960–1963. The objective of these experimental investigations was to obtain information on pressure jets burning in a furnace and to make use of the results both for the better understanding of basic combustion phenomena and also to interpret the experimental data in such a way that it would be directly useful for purposes of combustion engineering. The burners used in the trials consisted of a pressure

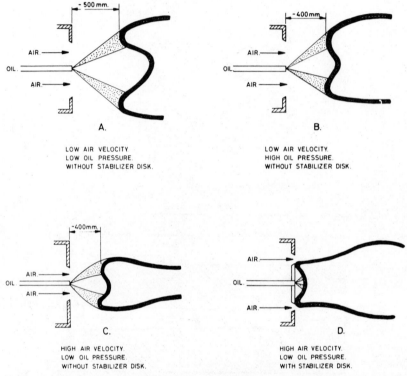

Fig. 6.15.   Pressure jet oil flames. Interaction of spray and air flows.

jet oil nozzle surrounded by an annular port for the introduction of combustion air. Simplex-type nozzles were used for the atomisation, to enable the independent variation of the oil atomising pressure and of the oil spray angle while maintaining the oil flow rate constant.

The experiments tested the effects of variation of air velocity, variation of oil pressure and the influence of introducing a stabiliser disc. The influence of these factors on the position of the flame front and the shape of the flame are summarised in Fig. 6.15.

In flames with low momentum flux of the air flow (burner exit velocity 10 m/sec and burner linear momentum 5 kg m/sec^2), the flame shape and the concentration and temperature distribution in the flame were determined more by the hollow cone pattern of the oil spray than by the aerodynamic pattern of the air flow. The flame front had a W-form with its edges inside the fuel-rich zones of the oil spray turned towards the burner. The flame front was stabilised at about 500 mm from the burner. An increase from the flame with low linear momentum flux to that with high momentum flux (burner exit velocity 70 m/sec and linear momentum flux 40 kg m/sec^2) was followed by a considerable change in flame characteristics. The flame front stabilised at about 400 mm from the burner, but the flame was much shorter and its width decreased so that it occupied only a small fraction of the total cross section of the furnace. The low momentum flame is bulky, highly luminous and of medium gas temperatures, while the flame with the high momentum flux has a small diameter, low emissivity and is a compact flame with high gas temperatures (peak temperature of 2023K) and a high volumetric heat release rate. Mixing patterns determined from detailed measurements of velocity, temperature and concentration within the flames show that the high fuel concentration zones coincide with the zones of high velocity gradients, where the shear stresses are high in the case of the high momentum flame but not in the case of the low air velocity flame. The volumetric heat release rates when calculated on the basis of flame volume, as given by the data of gas and soot concentration measurements, were of the order of $4 \cdot 65 \times 10^6$ W/m^3 in the high momentum flame compared with values smaller by more than an order of magnitude in the low momentum flame ($2 \cdot 9 \times 10^5$ W/m^3).

The fitting of a stabiliser disc at the burner mouth caused the flame to be stabilised at the burner when the oil spray angle was larger than 70°. However, when the spray angle was smaller, the vortex region behind the 200 mm diameter disc was not large enough to stabilise the flame on the disc because of the increased axial velocity component of the oil spray.

The effect of oil spray angle on the position of the flame front was considerable when no stabiliser disc was used. A small blue flame was stabilised in the central region of the wide stabiliser disc. The effect was observed while the air velocity was low. At conditions of higher air velocities, the influence of the oil spray angle was less important.

A detailed aerodynamic study was made of the flow region in the wake of the stabiliser disc under isothermal, non-burning conditions. This showed that approximately one-tenth of the momentum of the air was converted into static under pressure in the vortex region of the stabiliser disc. A ring vortex was found to be set up between the disc and the inside boundary of the annular jet. In the central region, reverse flow velocities of the order of one-twentieth of the exit velocity were found. Photographs of the flame near the burner exit have shown that the flow pattern determined under cold conditions still prevails when the oil spray is introduced at the centre of the disc and the fuel is ignited further downstream. The boundaries of the flow region downstream from the disc are marked on the photographs by fine oil droplets which are recirculated by the reverse flow at the stabiliser disc. These droplets move radially out from the axis towards the edge of the disc owing to the pressure gradient and are carried from there into the main annular jet stream. The presence of the oil drops on the face of the stabiliser disc caused heavy coke deposit formation.

In contrast with the flame front position, the combustion intensity was only slightly affected by the use of the stabiliser disc and by the variation of the oil spray angle. Under conditions in which the momentum flux of the air flow is increased, there was no significant effect of these variables upon the combustion intensity.

EFFECTS OF SWIRLING AIR FLOW

A number of experiments have been carried out at Ijmuiden[54] with pressure jet oil flames to examine the effects of swirling air flow upon the stability and combustion intensity of the flames. Air was introduced both tangentially and axially into the burners and by variation of these flow rates it was possible to vary the degree of swirl. Total oil and air flow rates, the oil atomising pressure and oil spray angle were maintained constant. A series of photographs taken of the flame front region of flames in the furnace are shown in Fig. 6.16. Each row of photographs is for a mean exit axial velocity of air which is maintained constant at 10, 30 and 50 m/sec respectively, while the ratio of the tangential to air flow rates increases from left to right. The flame represented by the photograph on the right-hand side of the middle row was produced with a wide angle oil spray, but it can be seen that the flame shape differs very little from that of the flame with 70° oil spray (third photograph from the left in the middle row). This demonstrates that when the momentum of the air flow is high, the variation of the oil spray angle has no significant influence on the flame properties. The photographs show that the variation of the degree of swirl gives effective control over the position of the flame front. It was found, however, that when the mean axial velocity (U) was low (10 m/sec) the degree of swirl

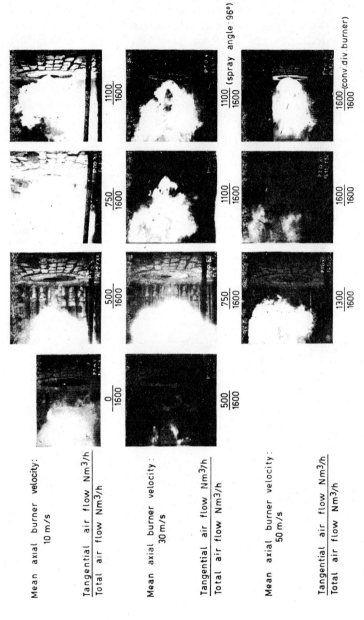

Fig. 6.16. Effect of variation of mean burner exit velocity and degree of swirl on pressure jet oil flames.

that was favourable for flame stabilisation could be excessive from the point of view of combustion efficiency. As a result of a too wide angle of spread of the swirling air jet, the air was directed outwards and was virtually separated from the fuel-rich zone of the jet. Thus mixing was impaired and combustion was incomplete. It was found that when U was increased to 30 m/sec, the value of the swirl number, $S = 0.78$, was sufficient to stabilise the flame at the burner and the mixing and combustion were improved giving an increased combustion intensity so that complete combustion could be obtained with about 0.5% excess oxygen. Planes of mixing ratios and degrees of oxidation showed that improved mixing resulted in a considerable decrease in flame volume when the linear and angular momentum were matched to produce an air jet in which the positions of the high velocity gradients (large shear stresses) coincided with the zones of high fuel concentrations.

When U was further increased to 50 m/sec, the flame could not be

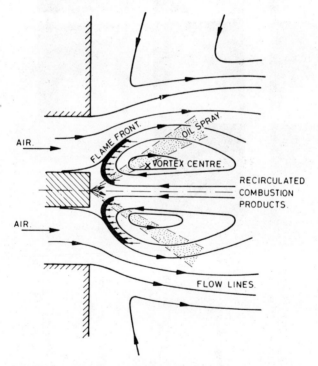

Fig. 6.17.   *Stabilisation of pressure jet oil flames by internal recirculation zone in swirling annular jet.*

stabilised at the burner even with the total air flow introduced tangentially into the burner (see bottom row in Fig. 6.16) except in the case when the annular port was extended with a short, water-cooled, divergent nozzle. In this case, the oil atomising nozzle was withdrawn into the throat of the convergent part of the burner extension. The flame front developed inside the water-cooled, divergent extension and complete combustion could be obtained with practically no excess air. The volumetric heat release rate calculated on the basis of the flame volume was of the order of 2.33×10^4 W/m^3.

Fig. 6.18.   *Fuel oil pressure jet spray flame in a furnace.*

Parallel with the furnace experiments, a detailed aerodynamic investigation was made of the swirling annular air jets used in the trials. Streamline air flow patterns as calculated from mean velocity measurements under isothermal conditions are shown in Fig. 6.17. Details from the photograph of the flame shown in Fig. 6.18 have been taken and superimposed upon the air flow patterns to give the composite picture shown in Fig. 6.17. This explains the mechanism of flame stabilisation whereby hot combustion products are recirculated within the central recirculation eddy and pass through the oil spray, forming a flame front at the position shown in the diagram. Small droplets are also carried into the centre of the hollow cone to form a 'laminarised' pilot flame. Further details giving the results of measurements within the flames and the influence of input variables are given in a series of publications and reports from the International Flame Research Foundation.[52-56]

REFERENCES

1. Dombrowski, N., *Biochemical and Biological Engineering Science*, Vol. 2, Ch. 16, Academic Press, London and New York, 1968.
2. Fraser, R. P., Proc. Joint Conference on Combustion, Inst. Mech. Engrs ASME Sec. 2, p. 76, 1955.
3. Rayleigh, Lord, *Theory of Sound*, 2nd Edn., Macmillan, London, 1894.
4. Weber, C., *A. angew Math. Mech.* 1931, **11**, p. 136.
5. Mellor, R., Chigier, N. and Beér, J. M., *Combustion and Heat Transfer in Gas Turbine Systems*, Cranfield International Symposium Series, No. 11, Pergamon Press, Oxford, 1971, pp. 291–305.
6. Hirschfelder, J. O. and Curtis, C. F., Third Symposium on Combustion, p. 121, Williams and Wilkins, Baltimore, 1949.
7. Godsave, G. A. E., Fourth Symposium on Combustion, p. 818, Williams and Wilkins, Baltimore, 1953.
8. Spalding, D. B., Fourth Symposium on Combustion, p. 847, Williams and Wilkins, Baltimore 1953.
9. Long, V. D., *J. Inst. Fuel* 1964, **37**, p. 522.
10. Goldsmith, M. and Penner, S. S., *Jet Propulsion* 1954, **24**, p. 245.
11. Wise, H., Lorrel, J. and Wood, B. J., Fifth Symposium on Combustion, p. 132, Reinhold, New York, 1955.
12. Hottel, H. C., Williams, G. C. and Simpson, H. C., Fifth Symposium on Combustion, p. 101, Reinhold, New York, 1955.
13. Okazaki, T. and Gomi, M., *J. Japan Soc. Mech. Engrs* 1953, **1–6**, p. 19.
14. Williams, F. A., *Combustion Theory*, Addison-Wesley, Massachusetts, 1965.
15. Kotake, S. and Okazaki, T., *Int. J. Heat and Mass Transfer* 1969, **12**, p. 595.
16. Peskin, R. L. and Wise, H., *AIAA* 1966, **4**(9), p. 1946.
17. Peskin, R. L., Polymeropoulos, C. C. and Yeh, P. S., *AIAA* 1967, **5**(12), p. 2173.
18. Polymeropoulos, C. C. and Peskin, R. L., *Combustion and Flame* 1969, **13**, p. 166.
19. Wise, H. and Ablow, C. M., *J. Chem. Phys.* 1957, **27**, p. 389.
20. Lorell, J., Wise, H. and Carr, R. E., *J. Chem. Phys.* 1958, **25**, p. 325.
21. Williams, F. A., *Combustion and Flame* 1961, **5**, p. 207.
22. Agoston, G. A., Wise, H. and Rosser, W. A., Sixth Symposium on Combustion, p. 708, Reinhold, New York, 1957.
23. Isoda, H. and Kumagai, S., Seventh Symposium on Combustion, p. 523, Butterworths, London, 1958.
24. Rosser, W. A., *Combustion and Flame* 1967, **11**(3), p. 442.
25. Graves, C. C. and Bahr, D. W., NACA Report 1300, p. 21, Ch. 1, 1957.
26. Frossling, N., *Gerlands Beitr, zur Geophysik* 1938, **52**, p. 170. AERE Harwell translation, Aug. 1963.

27. Ranz, W. E. and Marshall, W. R., *Chem Engr. Progr.* 1952, **48**, pp. 141, 173.
28. Eisenklam, P., Arunachalam, S. A. and Weston, J. A., Eleventh Symposium on Combustion, p. 715, The Combustion Institute, Pittsburgh, Pennsylvania, 1967.
29. How, M. E., *J. Inst. Fuel* 1966, **39**, p. 150.
30. Spalding, D. B., *Fuel* 1954, **33**, p. 255.
31. Nuruzzaman, A. S. M., Martin, G. F. and Hedley, A. B., 'Combustion of single droplets and simplified spray systems', *J. Inst. Fuel* 1971, **44**, No. 360, pp 38–54.
32. Spalding, D. B., *Some Fundamentals of Combustion*, Vol. 2, Butterworths Scientific Publications, London, 1955.
33. Eisenklam, P. and Arunachalam, S. A., *Combustion and Flame* 1966, **10**, p. 171.
34. Muggia, A., *L'Aerotecnica* 1956, **36**, p. 127. RAE Translation 679, Ministry of Supply, London.
35. Crowe, C. T., Nicholls, J. A. and Morrison, R. B., Ninth Symposium on Combustion, p. 395, Academic Press, London, 1963.
36. Williams, A., *Oxidation and Combustion Reviews* 1968, **3**(1), pp. 1–45.
37. Nuruzzaman, A. S. M., *Fuel Soc. J.* 1969, University of Sheffield, **20**, p. 30.
38. Goldsmith, M., *Jet Propulsion* 1956, **26**, p. 172.
39. Kumagai, S., Sixth Symposium on Combustion, p. 668, Chapman and Hall, London, 1956.
40. Masdin, E. G. and Thring, M. W., *J. Inst. Fuel* 1962, **35**, p. 251.
41. Kumagai, S. and Isoda, H., Fifth Symposium on Combustion, p. 129, Reinhold, New York, 1955.
42. Udelson, D. G., *Combustion and Flame* 1962, **6**, p. 93.
43 Godsave, G. A. E., NGTE Report R-125, 1952; NGTE Report R-126, 1953.
44. Topps, J. E. C., NGTE Memo. No. M-105, 1951.
45. Rex, J. F., Fuchs, A. E. and Penner, S. S., *Jet Propulsion* 1956, **26**, p. 179.
46. Kanevsky, J., *Jet Propulsion* 1956, **26**, p. 788.
47. Reichenbach, R., Squires, D. and Penner, S. S., Eighth Symposium on Combustion, p. 1068, Williams and Wilkins, Baltimore, 1962.
48. Iinuma, K., *Combustion and Flame* 1962, **6**, p. 127.
49. Nuruzzaman, A. S. M., Hedley, A. B. and Beér, J. M., 'Combustion of monosized droplet streams in stationary self supporting flames', Thirteenth Symposium on Combustion, Combustion Institute, pp. 787–99, The Combustion Institute, Pittsburgh, 1971.
50. Bolt, J. and Boyle, T. A., *Trans. ASME* 1956, **78**, p. 609.
51. Burgoyne, J. H. and Cohen, L., *Proc. Roy. Soc.* 1954, **A225**, p. 375.
52. The Joint Committee, International Flame Research Foundation (Holland), *J. Inst. Fuel* 1957, **30**, p. 553.
53. Beér, J. M., *Combustion* 1965, **37**(2), p. 27.
54. Beér, J. M., *Combustion* 1965, **37**(3), p. 41.
55. Beér, J. M. and Chigier, N. A., 'Swirling jet flames issuing from an annular burner', 5 me Journée d'Études sur les Flames, Paris, Nov. 1963.
56. Beér, J. M., *J. Inst. Fuel* 1962, **35**, p. 3.
57. Mellor, R., Ph.D. Thesis, University of Sheffield, 1969.

CHAPTER 7

Modelling of Combustion Systems

NOMENCLATURE

Ar	Archimedes number.
B	buoyancy force.
C	stoichiometric parameter (volume fraction of fuel in stoichiometric mixture); concentration.
c	specific heat.
D	diffusion coefficient.
D_s	diameter of the secondary burner [eqn. (7.13)].
d	nozzle diameter.
g	gravitational acceleration.
G	momentum flux.
K	scale factor (Table 7.1).
k	thermal conductivity.
L	length, flame length, half width of combustion chamber.
l	scale of turbulence.
M	mass flow rate.
p	pressure.
r	radial co-ordinate, radius.
T	temperature.
T_{sh}	turbulent shear force.
t	time.
U	velocity.
u'	turbulence intensity.
V	volume flow rate.
W	tangential velocity.
x, y, z	co-ordinates.

Greek Symbols

α	thermal diffusivity.
μ	viscosity.
v	kinematic viscosity.

ρ density.
θ Thring–Newby parameter.
τ shear stress.

Subscripts
A annular.
a ambient fluid.
dc double concentric.
e equivalent.
f fuel.
0   initial value.
s surrounding fluid, secondary fluid, source.

The ideal scientific method of establishing a law governing a physical or chemical process requires postulating a theory that rigorously describes the processes involved and subsequently carrying out an experiment to demonstrate the initially proposed hypothesis. Once the law has been established, predictions can then be made within the framework of the law without carrying out any further experiments. Combustion systems involving turbulent flames are too complex for such an ideal system to be used and, as an intermediate step, resort is made to modelling. Modelling connotes the practice of predicting the performance of a full scale plant by interpreting the results of a model experiment.

Use is made of (i) analogues in which the physical or chemical processes are of a different nature to those they simulate in the prototype, (ii) mathematical models in which experiments are carried out by mathematical analysis or numerical computations on a digital computer, and (iii) direct experimentation on physical models of the prototype.[1]

In the important range of practical investigations between the extreme cases of rigorous fundamental theory and prototype experiment, similarity considerations and model studies can often offer the quickest and least expensive route to the information required for development and design. The success of modelling depends largely on ensuring that there is similarity between the processes investigated in the model and in the prototype.

7.1 CATEGORIES OF SIMILARITY

Geometrical similarity is the simplest of model laws. It implies that every linear dimension of the model bears the ratio, $1/L$ to the corresponding dimension of the prototype. Although geometrical similarity can be maintained for studies of non-compressible fluid dynamics in isothermal systems, it has to be abandoned when the system is non-isothermal or when chemical reaction plays an important role in the process investigated. This then leads to the so-called 'distorted models'.

Mechanical similarity may be *static, kinematic* or *dynamic*. *Static* similarity requires that model and prototype undergo similar elastic or plastic deformations when corresponding stress systems are applied, while *kinematic* similarity in a system with flow implies that fluid or solid particles follow geometrically similar paths in corresponding intervals of time which is the same as saying that the 'streamline' pattern in model and prototype are geometrically similar. This also means that there is a velocity scale ratio which is maintained constant between corresponding points of model and prototype. The term 'streamline' is used here in a broad sense applicable also to turbulent flow. *Dynamic* similarity requires that the force ratios causing acceleration of masses in the corresponding systems are maintained constant.

Thermal similarity is concerned with systems which are non-isothermal and in which there is a flow of heat. In addition to kinematic similarity, which ought to be maintained if there is bulk movement of matter in the system, thermal similarity requires that the ratio of temperature differences between any two points in the prototype and corresponding points in the model are maintained constant.

7.2   METHODS OF ESTABLISHING SIMILARITY

DIMENSIONAL ANALYSIS

One of the general methods of establishing similarity is based on the Buckingham π-theorem.

Although the π-theorem can be employed most usefully for transforming a relationship of independent physical parameters into one of non-dimensional groups, the method has the disadvantage that it can be applied formally without requiring a better understanding of the reasons under-lying modelling laws. It can also lead to dimensionless groups which are unfamiliar or irrelevant, though certainly valid regarding requirements of dimensional analysis. The use of dimensional analysis in thermodynamics is greatly hampered also by the fact that the most important thermodyna-mic functions such as work, heat, enthalpy and energy have the same dimensions, and dimensional analysis does not distinguish between these.

Also there will be no difference in dimension between a differential quotient and a ratio of finite quantities.

It can be said that dimensional analysis is useful in that it provides a framework for the form of possible relationships but never the relationship itself.

The Derivation of Similarity from Fundamental Differential Equations

The fundamental relationships can best be introduced into the computations in the form of differential equations. These will usually be 'balance' equations expressing the conservation of mass or energy or the balance of forces. Equations expressing a balance of any of these entities will consist of terms that are dimensionally uniform and, by a simple division of a

Table 7.1

Dimensionless Groups obtainable from Differential Equations
(after Rohsenow and Choi[2])

Table 7.1A

Dimensionless Groups obtainable from Momentum Equation

Unsteady term	Inertia terms	Static pressure force	Viscous forces	Gravity force
$\dfrac{\partial U_z}{\partial t}$	$U_x \dfrac{\partial U_z}{\partial x} + U_y \dfrac{\partial U_z}{\partial y} + U_z \dfrac{\partial U_z}{\partial z}$	$\dfrac{1}{\rho}\dfrac{\partial p}{\partial z}$	$v\left(\dfrac{\partial^2 U_z}{\partial x^2} + \dfrac{\partial^2 U_z}{\partial y^2} + \dfrac{\partial^2 U_z}{\partial z^2}\right)$	g
$\dfrac{K_U}{K_t}$	$\dfrac{K^2_U}{K_L}$	$\dfrac{K_{\partial p/\partial z}}{K_\rho}$	$\dfrac{K_v K_U}{K^2_L}$	K_g

$$\frac{\text{Inertia}}{\text{Viscous}} \longleftarrow \frac{UL}{v} \longrightarrow$$

Reynolds number

$$\frac{\text{Inertia}}{\text{Gravity}} \longleftarrow \left\{\frac{U^2}{Lg}\right\} \longrightarrow$$

Froude number

$$\frac{\text{Static pressure}}{\text{Inertia}} \longleftarrow \left\{\frac{L(\partial p/\partial z)}{\rho U^2}\right\} \longrightarrow$$

Euler number

$$\frac{\text{Gravity}}{\text{Unsteady term}} \longleftarrow \frac{gt}{U} \longrightarrow$$

TABLE 7.1B

Dimensionless Groups obtainable from Energy Equation

Unsteady term	Convection terms	Conduction terms	Viscous terms
$\dfrac{\partial T}{\partial t}$	$U_x \dfrac{\partial T}{\partial x} + U_y \dfrac{\partial T}{\partial y} + U_z \dfrac{\partial T}{\partial z}$	$\alpha \left(\dfrac{\partial^2 T}{\partial x^2} + \dfrac{\partial^2 T}{\partial y^2} + \dfrac{\partial^2 T}{\partial z^2}\right)$	$\dfrac{\mu}{\rho c}\, \varphi$
$\dfrac{K_T}{K_t}$	$\dfrac{K_U K_T}{K_L}$	$\dfrac{K_\alpha K_T}{K^2_L}$	$\dfrac{K_\mu}{K_\rho K_c}\, \dfrac{K^2_U}{K^2_L}$

$\dfrac{\text{Conduction}}{\text{Unsteady term}} \longleftarrow\!\!\!\longrightarrow \dfrac{\alpha t}{L^2}$

Fourier number

$\dfrac{\text{Convection}}{\text{Conduction}} \longleftarrow\!\!\!\longrightarrow \dfrac{UL}{\alpha}$

Peclet number $=$ Re $\times$ Pr

$\dfrac{\text{Viscous}}{\text{Conduction}} \longleftarrow\!\!\!\longrightarrow \dfrac{\mu U^2}{kT}$

characteristic mass, energy or force, any of these equations can be transformed into a relationship between dimensionless groups. Such dimensionless groups have the advantage over those produced by the π-theorem that they are derived from the differential equations which describe quantitatively the important physical chemical processes in the system. Boundary conditions can then be used to reduce the number of dimensionless groups which include independent variables by introducing characteristic values

TABLE 7.1C

Dimensionless Groups obtainable from Mass-transfer Equation

Unsteady term	Convection terms	Diffusion terms
$\dfrac{\partial C}{\partial t}$	$U_x \dfrac{\partial C}{\partial x} + U_y \dfrac{\partial C}{\partial y}\, U_z \dfrac{\partial C}{\partial z}$	$D \left(\dfrac{\partial^2 C}{\partial x^2} + \dfrac{\partial^2 C}{\partial y^2} + \dfrac{\partial^2 C}{\partial z^2}\right)$
$\dfrac{K_c}{K_t}$	$\dfrac{K_U K_c}{K_L}$	$\dfrac{K_D K_c}{K^2_L}$

$\dfrac{\text{Diffusion}}{\text{Unsteady term}} \longleftarrow\!\!\!\longrightarrow \dfrac{Dt}{L^2}$

$\dfrac{\text{Convection}}{\text{Diffusion}} \longleftarrow\!\!\!\longrightarrow \dfrac{UL}{D}$

Re $\times$ Sc

instead of length, velocity, etc. (*e.g.* nozzle diameter and exit velocity from the nozzle).

Table 7.1 represents the derivation of some of the important dimensionless groups from differential equations of momentum, heat and mass transfer.

The equation at the top of Table 7.1a is the special form of the Navier–Stokes equation for incompressible fluid and constant density, viscosity and temperature. The second line gives the 'scale' coefficients according to dimensional analysis procedures. The following lines contain some dimensionless groups formed by ratios of forces. Table 7.1b follows the same procedure starting from the energy equation and Table 7.1c from the mass-transfer equation. The tables are in the form as given by Rohsenow and Choi.[2]

The mere production of dimensionless groups which must be held constant in prototype and model is, however, only the first step towards successful modelling. In most cases it is not possible and not necessary to aim at complete modelling, *i.e.* to reproduce all prototype processes in the model, and it is the central task of the scientist or engineer to select critically those groups which need to be retained for successful modelling of certain aspects of the prototype system. This critical selection requires insight and a good knowledge of the processes involved and is often referred to as the art of partial modelling.

SIMILARITY IN CHEMICAL REACTORS

Damköhler[3] was perhaps the first to formulate model laws for chemical reactors. He deduced that similarity between model and prototype required the maintenance of five dimensionless ratios. These are:

I.   Chemical reaction/Bulk flow;
II.   Chemical reaction/Molecular diffusion;
III. Heat liberated/Heat transported by bulk flow;
IV.   Heat liberated/Heat transported by conduction;
V.   Momentum transferred by bulk flow/Momentum transferred by viscosity (Re number).

It is usually impossible to maintain all five groups constant, *e.g.* for a given reaction the equality of Reynolds numbers is incompatible with the requirement for equal residence times in model and prototype.

It is because of the difficulties of observing all similarity conditions in modelling that there is great interest in partial modelling.

7.3   SIMILARITY RULES THAT CAN BE NEGLECTED

It is not possible to give hard and fast rules as to which similarity rules can be relaxed and which have to be observed. This depends greatly on the specific question for which an answer is required. There are, however, general cases of combustion modelling in which the experimenter can safely ignore some of the rules. For example:

(a) If the Reynolds number is sufficiently large and the process investigated is controlled by turbulent mixing, the molecular transport processes in such flows can be neglected and there is therefore no need to maintain equal Prandtl or Schmidt numbers in model and prototype.

$Re >> 2300$

(b) If the Reynolds number is sufficiently large and interest is in a fully separated flow (jet or wake flow) in the enclosure there is no need to maintain the same Reynolds number in model and prototype. An example of this would be the investigation of the region of a turbulent jet flame before the jet expands to fill the cross section of the combustor.

$Re >> 20,000$

(c) If the Froude number is sufficiently high, *i.e.* the inertial forces are large compared with buoyancy effects, there is no need to maintain the same value of this dimensionless group in model and prototype. It was shown, however, in the case of a certain industrial furnace with low burner velocities, that the buoyancy effects cannot be neglected.[4]

$Fr = \dfrac{U^2}{Lg}$

(d) Two-phase flow effects need not be simulated provided that the settling velocity of oil drops or of a cloud of particles is not commensurable with the bulk flow velocity.

$U_t << U$

(e) When, however, the burning of a cloud of droplets is modelled, the droplet size should be scaled in relation to the ratio of length scale to velocity scale, *i.e.* in proportion to the residence time. This follows from the fact that the burning rate of a droplet is proportional to its diameter.

$d \propto \dfrac{L}{U} = t$

(f) When non-isothermal conditions in the prototype are simulated with isothermal flow in the model, the geometrical similarity ought to be abandoned in order to maintain the mass flow and velocity

$d_e = d_0 \left(\dfrac{\rho_0}{\rho_h} \right)^{1/2}$

ratios while distorting the nozzle diameter in the model by the ratio of the square root of the densities of the cold fluid and of the hot flame gases. This modelling principle was first advocated by Thring and Newby[6] and has been successfully used by many investigators.

7.4 APPLICATION OF PARTIAL MODELLING TO FLAMES

The principle of partial modelling has been successfully applied to diffusion flames—the most common type of industrial flame. In these flames the mixing of the reactants and of hot, fully burned combustion products is the process that determines the overall rate of combustion and flame structure because the rate of chemical reaction is high at flame temperatures compared with the rate of mixing. Mixing in turn depends on the pattern of flow and further on the forces that are operative in producing the flow pattern. In the following, dimensionless modelling criteria are developed by forming ratios of force terms, the latter of which are chosen according to special physical conditions in the flame. This discussion follows a line of argument presented by Hottel.[5]

The forces to be considered for diffusion flames include:

(1) Momentum flux at source: (kg/sec) $\times$ (velocity) $= MU = G_s$.
(2) Viscous shear forces: (area) $\times$ (velocity gradient perpendicular to the area) $\times$ (viscosity), where the viscosity μ is proportional to (mean free path $\times$ mean molecular velocity $\times$ density).
(3) Turbulent shear force: (area) $\times$ (velocity gradient perpendicular to the area) $\times$ (turbulent or eddy viscosity), where the eddy viscosity μ_e is (scale of turbulence $\times$ turbulent velocity $\times$ density). Near the nozzle in a jet flow, the scale is proportional to the nozzle diameter, further downstream the scale becomes proportional to the dimensions of the enclosure.
(4) Buoyancy force: (volume) $\times$ (density difference) $\times$ (gravitational acceleration) $=$ volume $(\rho_{amb} - \rho_{flame})\, g$.
(5) Centrifugal force: (volume) $\times$ (density) $\times$ (centrifugal acceleration) $=$ volume $\times \rho \times W^2/r$.

Similarity between model and prototype can be assured by maintaining the value of the characteristic force ratios at corresponding points of model and prototype.

Hottel[5] discusses three cases of practical interest:

(a) An open turbulent jet flame with the source momentum and turbulent shear as the significant forces,
(b) A buoyant, turbulent flame with the turbulent shear force and the buoyancy force as significant forces, and
(c) A buoyant, turbulent jet flame with the turbulent shear force, the buoyancy force and the source momentum flux as the significant forces.

For case (a) the dimensionless force ratio to be maintained at corresponding points in the prototype and in the model flame is that of the source momentum flux to the turbulent shear force. For the turbulent shear force we can write

$$T_{sh} \sim L^2 \frac{dU}{dy} lu'\rho \sim L^2 U^2 \rho \tag{7.1}$$

This is because both the velocity gradient and the intensity of turbulence are proportional to the linear velocity U. The linear velocity at the end of the flame can be given as

$$U \sim \frac{V}{C} \left(\frac{\rho_{amb}}{\rho_{flame}}\right) \left(\frac{1}{L^2}\right) \tag{7.2}$$

where V is the volume flow rate in m^3/sec, C is the stoichiometric parameter, *i.e.* the volume fraction of fuel in a stoichiometric mixture with air (*e.g.* $C = 1/10\cdot5$ for a methane–air flame), and L is the flame length.

Substituting for U in eqn. (7.1) we have

$$T_{sh} \sim \frac{V^2}{C^2} \left(\frac{\rho_{amb}}{\rho_{flame}}\right)^2 \frac{\rho}{L^2} \tag{7.1a}$$

On the other hand, the source momentum flux

$$G_s \sim \frac{\rho_0 V^2}{d^2} \tag{7.3}$$

so the ratio

$$\frac{G_s}{T_{sh}} = \frac{\rho_0 V^2}{d^2} \frac{C^2 \rho_f{}^2 L^2}{V^2 \rho_a{}^2 \rho_f} = \frac{L^2}{d^2} C^2 \left(\frac{\rho_f \rho_0}{\rho_a{}^2}\right) \tag{7.4}$$

For case (b) the dimensionless force ratio will be the ratio of turbulent shear and buoyancy forces. The turbulent shear force

$$T_{sh} \sim L^2 U^2 \rho \sim \left(\frac{V}{C}\right)^2 \left(\frac{\rho_a}{\rho_f}\right)^2 \left(\frac{\rho_f}{L^2}\right)$$

The buoyancy force

$$B \sim L^3(\rho_{amb} - \rho_f)g$$

and the dimensionless force ratio is then

$$\frac{T_{sh}}{B} \sim \left(\frac{V}{C}\right)^2 \frac{1}{L^5} \frac{\rho_a^{\,2}}{\rho_f(\Delta\rho)g} \tag{7.5}$$

For case (c), where the flame characteristics depend on three forces (source momentum flux, turbulent shear force and buoyancy force), Hottel[5] suggests that one of the dimensionless force ratios developed for case (a) might be given as a function of the other dimensionless group developed for case (b) as

$$\left(\frac{LC}{d}\right) \frac{(\rho_f\rho_0)^{1/2}}{\rho_a} = f[(gd^5\rho_a^{\,3}\Delta\rho)/(V^2C^3\rho_f^{\,3/2}\rho_0^{\,5/2})] \tag{7.6}$$

By expressing L from the first force ratio eqn. (7.4) and substituting it into the second force ratio eqn. (7.5), we obtain the right-hand side of the above equation which is a modified Froude number or Archimedes number. This can be seen by substituting d^2U for V in the denominator of the right-hand side.

The Free Turbulent Jet

Another method of deriving modelling criteria is based on the unique characteristic of turbulent jets, *i.e.* all fully turbulent, constant density, subsonic, free jets are similar. Most practical combustion systems consist of a complex arrangement of turbulent jets, but there is a good possibility of leading back the problem to the basic case of the free, constant density, turbulent jet.

Non-isothermal Jets

When the temperature of the jet fluid T_0 is higher than that of the surrounding fluid T_s the rate of decrease of the velocity at the jet axis with distance from nozzle is greater than it is for isothermal jets.[6] The curves can be made to coincide with that for an isothermal constant density jet if the orifice diameter d_0 is replaced by an 'equivalent diameter':

$$d_e = d_0 \left(\frac{T_s}{T_0}\right)^{1/2} = d_0 \left(\frac{\rho_0}{\rho_s}\right)^{1/2} \tag{7.7}$$

This concept of the equivalent diameter is based on the assumption that the whole development of the jet is determined by the total jet momentum.

The total momentum flux for a single jet can be written as

$$G = 2\pi \int_0^\infty \rho U^2 r \, dr = G_0 = \frac{\pi}{4} d_0{}^2 \rho_0 U_0{}^2 = \frac{\pi}{4} d_e{}^2 \rho_s U_0{}^2 \qquad (7.8)$$

and hence

$$d_e = d \left(\frac{\rho_0}{\rho_s}\right)^{1/2}$$

This is then the nozzle through which the fluid would flow with the same mass flow rate and momentum, but with the density of the surrounding fluid instead of the nozzle-fluid density.

When the mass flow rate through the nozzle M_0 is known and the jet momentum flux G_0 is measured, the equivalent nozzle diameter can be given as

$$d_e = \frac{2M_0}{(G_0 \pi \rho_F)^{1/2}} \qquad (7.7a)$$

This formula can be used, for example, for modelling the mixing in a steam-atomised oil jet. The density in the formula ρ_F is that of the combustion products at flame temperature in the burning jet.

ENCLOSED JETS

For enclosed jets, such as a jet issuing from a nozzle into a duct, two extreme cases can be considered:

  (a) There is ample supply of secondary fluid surrounding the jet so that the jet can freely entrain until it expands to reach the wall of the duct.
  (b) The surrounding secondary fluid is less than that which the jet can entrain.

In case (b) a recirculation flow will be set up downstream of the point where all secondary fluid has been entrained. The mass concentration of nozzle fluid after complete mixing in the duct can be given as

$$C_\infty = \frac{M_0}{M_s + M_0} \qquad (7.9)$$

where M_0 is the mass flow rate of nozzle fluid and M_s is the mass flow rate of secondary fluid (usually air). It can be shown that $C_\infty L/r_0$ must be equal in model and prototype to ensure similarity, where r_0 is the nozzle radius and L is the half width of the duct.[6]

When combustion is occurring in an enclosed jet, similarity can be ensured by using the equivalent nozzle diameter concept mentioned in the previous paragraph combined with the above similarity criterion to give

$$\left(\frac{M_s + M_0}{M_0}\frac{r_e}{L}\right)_{\text{model}} = \left(\frac{M_s + M_0}{(G_0\pi\rho_F)^{1/2}}\frac{1}{L}\right)_{\text{hot system}} \quad (7.10)$$

The recirculation parameter

$$\theta = \frac{M_s + M_0}{M_0}\frac{r_e}{L} \quad (7.11)$$

proposed by Thring and Newby[6] enables quantitative predictions of the flow pattern in the duct to be made, such as the mass flow rate of recirculated flow along the duct, the position of the core of the recirculation eddy, etc. Experimental evidence has been collected covering the values of θ from 0·02 to 0·6;[7] industrial practice generally lies between 0·15 and 0·30.

DOUBLE CONCENTRIC JETS

Double concentric jets are frequently used as a burner configuration with the fuel jet as the central jet surrounded by the annular air jet. In double concentric jets, both the primary and secondary jets have considerable momentum and the secondary nozzle is usually small compared with the width of the furnace: $r_2 \leq L$. Analyses of double concentric jets showed that, some distance downstream from the nozzle, the primary and secondary streams combine and the velocity distribution can be described at that point as a single Gaussian distribution.[8] It was also shown that for these conditions the Thring–Newby recirculation parameter [left-hand side of eqn. (7.10)] reduces to

$$\theta = \frac{(r_e)_{dc}}{L} \quad (7.11a)$$

as $M_s = 0$, where the equivalent nozzle radius can be calculated from the combined masses and momenta of the primary and secondary flows as

$$(r_e)_{dc} = (M_0 + M_s)/[(G_0 + G_s)]\pi\rho_F]^{1/2} \quad (7.7b)$$

When similarity is required in the region near the nozzle where the two streams have not yet combined, it is necessary to maintain the ratio of the input mass flows of the primary to secondary streams the same in the model and the furnace:[9]

$$\left(\frac{M_0}{M_s}\right)_{\text{model}} = \left(\frac{M_0}{M_s}\right)_{\text{hot system}} \quad (7.12)$$

BUOYANT JETS

In certain cases such as cement kilns and glass tanks, where the combustion air is supplied around a primary burner with low velocity (2 m/sec), the ratio of gravitational to inertia forces is too high and not only is the path of the jet axis deflected, but the mixing between the jet and its surroundings is affected.

Under these conditions rule (c) (p. 202) cannot be relaxed and similarity requires that the Archimedes number—which is a modified Froude number for non-constant density jet flow systems—is maintained the same in model and prototype.

The Archimedes number can be given for such cases in the form

$$Ar = \frac{D_s g \, \Delta T}{U_s^2 T_g} \tag{7.13}$$

where D_s is the diameter of the secondary burner, $\Delta T = T_g - T_s$ (where T_g is the temperature of the fully burnt gases in the furnace, T_s is the temperature of the secondary air) and U_s is the secondary air velocity.

It was shown[9] that the effect of buoyancy becomes significant when $Ar > 0.01$. The predicted path of the axis of the jet deflected by buoyancy can be given as

$$\frac{y}{D_s} = 0.065 \, Ar \left(\frac{x}{D_s}\right)^2 \tag{7.14}$$

which was shown to be in general agreement with measurements made in the Ijmuiden furnace.

Swirling jets having similar swirl generators can be characterised by the non-dimensional ratio of the angular to linear momentum flux:[10]

$$\frac{G_\varphi}{G_0 r_A} = \frac{2\pi \int_0^\infty \rho U W r^2 \, dr}{r_A(2\pi \int_0^\infty (\rho U^2 + P) r \, dr]} \tag{7.15}$$

where G_φ is the angular momentum flux, G_0 is the linear momentum flux, r_A is the outer radius of the annulus, U is the axial velocity component, W is the tangential velocity component, $P = p - p_\infty$ (static pressure–ambient static pressure) and r is the radial distance from jet axis.

7.5   RESIDENCE TIME DISTRIBUTIONS IN COMBUSTORS

The problem of residence time distribution (rtd) in continuous flow systems is of great interest mainly because of its potential in predicting the performance and efficiency of reactors. Two particular types of flow can be considered as extreme cases from the point of view of rtd: the 'well stirred' flow, where the properties of the fluid in the vessel are uniform and identical with those of the outgoing stream, and 'plug flow', in which elements of fluid move through the reaction vessel with constant and equal velocity. In the former type of flow the rtd can be shown to be of the form

$$\psi(t) = \bar{t}^{-1} \exp\left(-t/\bar{t}\right) \qquad (7.16)$$

where t is time and $\bar{t}$ is mean residence time (reactor volume divided by volume flow rate of fluid). In a plug flow system, however, all fluid particles have residence times equal to the mean residence time $\bar{t}$. It is evident that a distribution of residence times is a consequence of departure from plug flow, and that the degree of this departure can also be characterised by the recirculation pattern. It was shown,[11] in a study carried out with swirling jets both in a 1/10th scale cold model of the Ijmuiden furnace and in the furnace itself while burning pulverised coal, that rtd can be modelled in an isothermal system by using the equivalent nozzle diameter concept together with maintaining the same 'degree of swirl' in model and prototype.

7.6 EXPERIMENTAL METHODS

Isothermal modelling rules enable the use of cold water or air models. This simplifies both the construction of models and also the experiments.

Flow visualisation often provides the shortest route for the engineer to the solution of his problem. It can also be invaluable as a preliminary study before a detailed quantitative experimental investigation is undertaken. Fine balsa dust or smoke can be used as tracer in the air models and polystyrene particles, air bubbles or aluminium powder in the water models. The flow is made visible in a cross section of the transparent plastic model by a high-intensity, narrow plane beam of light cutting a slice across the model, and can be viewed or photographed at right angles to the plane beam.

Water models and the above visualisation technique were widely used by combustion workers in connection with research on open-hearth furnaces.[12] The application then spread to gas turbine combustor studies[13] making use of isothermal modelling laws.

Density changes due to combustion are in certain cases not negligible. Density variations in a liquid-fuel rocket combustor were simulated for flow visualisation by the interesting technique of the flash vaporisation of liquid kerosine.[14]

Although flow visualisation studies may sometimes enable limited quantitative information to be obtained, *e.g.* trace lengths on photographs can be interpreted to determine local flow velocities, the results are mainly qualitative.

Quantitative model investigations of velocity, static pressure and concentration distributions require probe measurements to be made. Tracers are used for the study of mixing of jet flows and surrounding fluid and the mixing is determined from the analysis of samples withdrawn from various points in the model. The tracer can be temperature (*e.g.* the jet fluid can be preheated) and local 'mixedness' can then be determined from temperature measurements. In three-component mixing, such as occurs in double concentric enclosed jets between primary fluid, secondary fluid and recirculation, two tracers ought to be used, one of which can be temperature and the other concentration. In a detailed study of this kind the primary air was preheated to 80°C and the recirculation was labelled with CO_2. The local concentrations in terms of primary, secondary fluid and recirculation were then determined from measurements of temperature and CO_2 concentration analyses of the samples.[9]

'Transient' tracer techniques can be used conveniently for residence time distribution studies both in isothermal models and in prototypes. In a series of experiments at Ijmuiden[11] a salt solution was introduced into a water model and, after steady-state condition was reached, the tracer introduction was interrupted and the decay of the tracer concentration with time was measured in the exit stream. Essentially the same technique, but with helium as tracer, was used for residence time distribution measurements in the prototype furnace.[11]

REFERENCES

1. Beér, J. M., *J. Inst. Fuel* 1966, **39**, p. 466.
2. Rohsenow, W. M. and Choi, H. Y., *Heat, Mass and Momentum Transfer*, Prentice Hall, New Jersey, 1965.
3. Damköhler, G., *Der Chemie-Ingenieur* (A. Eucken and M. Jakob, eds), Band III, Teil 1, Akademische Verlag, 1937.
4. Chigier, N. and Beér, J. M., Int. Flame Res. Found. Doc. No. GO2/a/4, Ijmuiden, 1963.
5. Hottel, H. C., Unpublished notes on the technical session, Amer. Flame Research Comm., 1965.

6. Thring, M. W. and Newby, M. P., Fourth Symposium on Combustion, pp. 789–96, Williams and Wilkins, Baltimore, 1953.
7. Barchilon, M. and Curtet, R. J., *J. Basic Eng.* (*Trans. ASME*) 1964, p. 777.
8. Chigier, N. A. and Beér, J. M., *J. Basic Eng.* (*Trans. ASME*) 1964, pp. 797–804.
9. Beér, J. M., Chigier, N. and Lee, K. B., Ninth Int. Symposium on Combustion, pp. 892–900, Cornell University, N.Y., 1962.
10. Chigier, N. and Beér, J. M., *J. Basic Eng.* (*Trans. ASME*) 1964, pp. 788–96.
11. Beér, J. M. and Lee, K. B., Tenth Symposium on Combustion, pp. 1187–1202, The Combustion Institute, 1965.
12. Chesters, J. H., Howes, R. S., Halliday, I. M. D. and Philip, A. R., *J. Iron Steel Inst.* 1949, **162,** p. 385.
13. Clarke, A. E., Gerrard, A. J. and Holliday, L. A., Ninth Symposium on Combustion, p. 878, Cornell University, N.Y., 1962.
14. Herne and Thring, M. W., Ninth Symposium on Combustion, p. 965, Cornell University, N.Y., 1962.

BIBLIOGRAPHY

Johnstone, R. E. and Thring, M. W., *Pilot Plants, Models and Scale-up Methods in Chemical Engineering*, McGraw-Hill, New York, 1957.

Hansen, A. G., *Similarity Analyses*, Prentice Hall, New Jersey, 1965.

Klinkenberg, A. and Mooy, H. H., *Chem. Eng. Prog.* 1948, **44,** p. 17.

Spalding, D. B., Ninth Symposium on Combustion, pp. 833–43, Cornell University, N.Y., 1962.

Traustel, S., *Modellgesetze der Vergasung und Verhüttung*, Akademie Verlag, 1949.

CHAPTER 8

Measurements in Flames

NOMENCLATURE

A, B, C	calibration constants for the hot wire.
d	diameter of wire sensor on probe.
D	diameter of swirler.
E	probe output voltage.
G, K	probe direction constants.
L	length of hot wire sensor.
r	radius.
T	time.
U	mean resolved velocity acting on hot wire sensor.
u, U	velocity in direction u.
v, V	velocity in direction v.
w, W	velocity in direction w.
x	axial co-ordinate.

Greek Symbols

α	phase difference.
ρ	density.
ψ	stream function.
ψ_0	non-dimensional stream function.

Subscripts

U, V, W	co-ordinate system directions.
1, 2, 3, 4	measuring planes of hot wire probes.
max, min	maximum or minimum velocity measured with the assumed velocity.

Superscripts

$-$	time mean average values.
$'$	fluctuating values.
$\sim$	square root of fluctuating quantity squared, i.e. $(\overline{u'^2})^{\frac{1}{2}}$.

212

8.1 INTRODUCTION

Measurements in flames are made with the main objective of establishing the rates of transfer of heat, mass and momentum within the flame and from the flame to surfaces outside the flame. Measurements are made of the spatial distributions of velocity; pressure; temperature; concentration of gases and solid or liquid particles; convective and radiative heat transfer; turbulence and electrical properties. In turbulent systems, it is necessary to measure 'instantaneous' values which can be separated into time mean average values, root mean square values and time average correlations for determination of local values of mass, heat and momentum flux.

Prediction methods allow, in principle, the determination of the above mentioned quantities by solution of the relevant equations as a function of the initial and boundary conditions. Despite considerable progress in the use of numerical methods for solution of the equation,[1] knowledge of flame systems is mainly dependent upon direct measurement. Further, the development of prediction methods requires detailed measurement to be made for the purposes of testing theories and hypotheses of turbulent transport of mass, heat and momentum. Measurement is also required for determining accurate physical input data and quantities such as radiative properties of gaseous compounds and particles in flames.

Most engineering systems are too complex for the application of prediction methods and, at present, knowledge of the systems is almost entirely dependent upon experimental studies. Progress is being made in the development of similarity and modelling laws which allow predictions to be made on the basis of measurements in a similar system or from comparative studies in isothermal models on the one hand and pilot plants or industrial scale prototypes on the other.

It is frequently found in research and development work that systematic experimental studies, even with only a limited possibility of generalisation of results,can provide useful data for specific engineering problems.

Principles applied in instruments for measurements in flames are described in a number of books[2-5] which also give details of design and construction of instruments for use in laboratories. In the last two decades a considerable effort has been made by investigators at the International Flame Research Foundation, Ijmuiden, Holland, to develop measuring techniques for the study of flames in industrial size combustion chambers and furnaces. Reference to this work can be found in a series of books, papers and reports.[6-12]

Measurement techniques and instruments are discussed in the following groups:

> gas velocity
> pressure;
> temperature;
> heat transfer;
> gas and solid sampling;
> drop size and trajectories in sprays;
> turbulence in flames.

8.2   GAS VELOCITY

Local gas velocity is one of the primary aerodynamic variables which characterise a flame, and the distribution of velocity throughout the system is one of the first requirements for a complete description of the flow system. The rate of flame propagation normal to the flame front, generally referred to as the flame speed, is determined by measuring the velocity of gases entering normal to a stationary flame front. Streamlines of gas flow, mass flow rates within stream tubes or annuli, mass flow rates within recirculation eddies and total entrained mass flow rates can all be calculated from integration of measured mean velocity distributions.

The most direct means of measuring velocity is the *particle track method* in which measurement is made of the time taken for a particle to move across a known or measured distance. Fristrom and Westenberg[2] describe a photographic system for measurement of velocity in flames using magnesium oxide particles as a tracer and intermittent illumination for photographing the particle tracks. The method has been used mainly in laminar flame studies but is also applicable to turbulent flames.

The other methods of determining velocity are indirect and involve insertion of probes into the flow. The pitot tube method involves transfer of kinetic energy into pressure energy, while the hot wire anemometer involves measuring the rate of heat transfer to a wire as it is affected by changes in local gas velocity.

Pitot Tube
The standard pitot tube is made up of a central tube in which the total pressure is measured by the impact of the flow onto the tube facing the oncoming stream. The static pressure is measured through circumferential holes at right angles to the flow direction. The velocity is calculated from

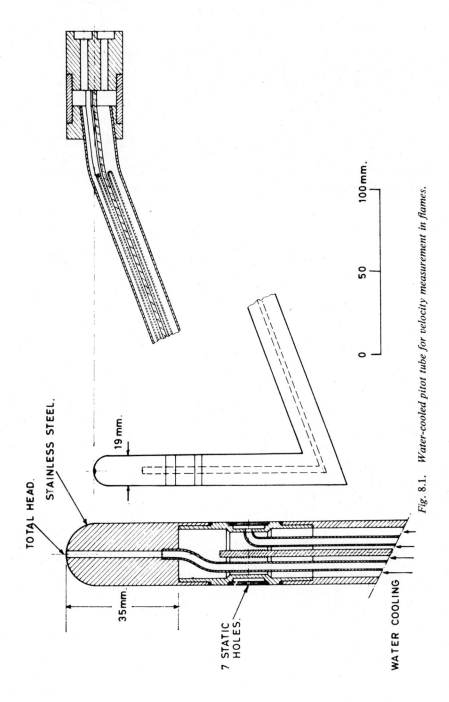

TOTAL HEAD.

STAINLESS STEEL.

19 mm.

35 mm.

7 STATIC HOLES.

WATER COOLING

0   50   100 mm.

Fig. 8.1.   Water-cooled pitot tube for velocity measurement in flames.

the measurement of the differential pressure and from the density of the gases according to the equation.

$$V = \left(\frac{2(P_t - P_s)}{\rho}\right)^{1/2}$$

where $P_t - P_s$ is the pressure difference between total and static pressure measurements and ρ is the density of gases at the measuring point. The local density must be determined from separate local measurements of temperature and concentration of species using the equation

$$\rho = \frac{P\bar{M}}{RT}$$

where P is pressure (atm); $\bar{M}$ is average molecular weight (g/mole); T is temperature (°K); and R is the molar gas constant (82·05 cm^3 atm/mole/°K).

For small flames with relatively low temperatures, small probes may be constructed from quartz or stainless steel. For measurements in furnaces under conditions of high temperature, with the presence of dust and liquid particles, a more robust instrument must be used in order to prevent damage and blockage. A water-cooled pitot with a stainless steel head is shown in Fig. 8.1.[8] The stainless steel tip, which is uncooled, is heated by the flame and in oil flames blockage of the holes is prevented or considerably reduced owing to burning or vaporisation of oil at the probe tip. The head is bent so as to be at right angles to the stem while the probe end is in line with the axis of the stem.

FIVE-HOLE PITOT

In systems with three-dimensional flow, such as flows with swirl or those involving recirculation, it is necessary to measure both the magnitude and the direction of the velocity. The use of standard yaw and pitch meters requires rotation of the probe and in many systems this is not possible or it is inconvenient. Measurement of velocity magnitude and direction, as well as the static pressure, can be made with a five-hole pitot. The principle of the instrument is based upon the measured pressure distribution around a sphere introduced into a flowing stream. Lee and Ash[13] have shown that, with five holes on the circumference of a sphere, the instrument can be calibrated so that, from a measurement of three differential pressures, the magnitude and direction of the velocity vector can be determined.

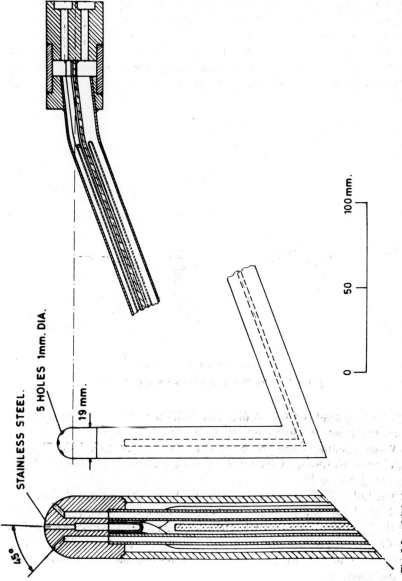

STAINLESS STEEL.

5 HOLES 1mm. DIA.

19 mm.

45°

0   50   100 mm.

Fig. 8.2.   Multidirectional impact tube (water cooled) for measurement of magnitude and direction of velocity in flames.

In order to reduce the dimensions of the sensing head to a minimum and still allow for water cooling of the probe, a hemispherical head is recommended as shown in Fig. 8.2.[8] Five holes are drilled into the stainless steel head—a central hole surrounded by four others at an angle of 45° to the axis. The holes are connected to the five pressure tubes in the water-cooled section of the head which is 19 mm in diameter. For cold flow studies the small five-hole pitot tube shown in Fig. 8.3 can be used.[8]

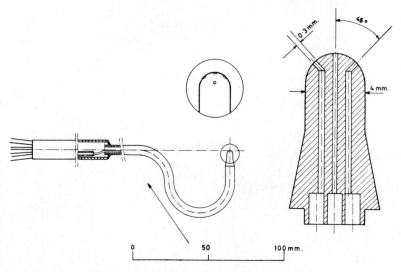

Fig. 8.3. *Multidirectional impact tube (uncooled).*

The sensing head is 4 mm in diameter with the holes 0·3 mm in diameter. This instrument is particularly useful in regions of large velocity gradient.

Water cooling of the probe head can be avoided by constructing the head from platinum. A platinum five-hole pitot has been used in flames by Macpharlane.[14]

The holes of pitot probes become blocked with particles of dust, soot and liquid fuel and provision needs to be made for frequent blowing out of the holes with nitrogen. Calibration of the five-hole pitot tubes can be made in the potential core region of an air jet and is carried out by progressively changing the yaw and pitch angles and measuring three differential pressures.

Pitot tubes have many limitations. Probe sizes are large and cause considerable disturbance to the system. Velocity levels in flames are often very small and because of the low density of gases it is usually required to measure pressures of less than 1 mm water gauge. In recirculation zones, static pressures are sub-atmospheric and a separate traverse is necessary to measure static pressure. Little information is available about the effect of turbulence intensities on pitot tubes, but the effect is undoubtedly considerable for turbulence intensities higher than 20%. An indication of the degree of error is given by the expression

$$p_m = p_{si} + c_i \rho \overline{u'_s}^2$$

where p_m is measured pressure; p_{si} is the true pressure; $\overline{u'^2}$ is the fluctuating mean velocity component; and the constant c_i varies between 0 and $\frac{1}{4}$.

There are three such equations and if the turbulence intensity in all three directions is larger than 20% these simple expressions do not hold. The procedure of making corrections for turbulence levels is long, involved and tedious. In general, the frequency response and sensitivity of pitot tubes at low flow velocities is low.

HOT WIRE ANEMOMETERS

The hot wire anemometer has the necessary high frequency response and high sensitivity at low flow velocities for measurements in the high turbulence levels encountered in three-dimensional flow systems. Its use provides a means of evaluating mean and fluctuating velocity components in highly turbulent flow fields. The method has been extensively evaluated and has been found to be more accurate than previous methods developed.

Hot wire anemometry is a well known fluid velocity measurement technique.[15-17] A typical hot wire is shown in Fig. 8.4.

The basic measurement is of heat loss from a wire or film to a fluid stream. Usually this heat loss is interpreted in terms of velocity for a constant temperature, constant pressure air stream. However, the hot wire is also sensitive to temperature, density and composition fluctuations.

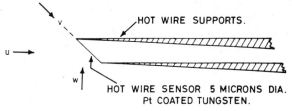

Fig. 8.4. Hot wire probe and velocity components.

For high velocity flows, density changes become an important parameter in the heat loss equation. However, only isothermal, incompressible fluid flow is considered here.

The hot wire anemometer may be used in either of two modes:

(a) the constant current mode;

(b) the constant temperature or constant resistance mode.

For most systems (b) has superseded (a). In a constant current mode, considerable temperature fluctuation can occur on the hot wire and supports causing calibration errors and loss of frequency response due to thermal inertia. The constant temperature system minimises the thermal inertia of the probe by operating at a constant resistance (temperature) and using the heating current as a measure of heat transfer and hence velocity.

The anemometer consists basically of a Wheatstone bridge, of which the probe forms part, and a D.C. amplifier. The probe temperature is kept constant by keeping the bridge in balance and so keeping the temperature-dependent probe resistance constant. Current fed to the bridge top from the amplifier serves the two-fold purpose of indicating bridge balance conditions and heating the probe—the amplifier being polarised so that a change in probe resistance due to cooling will result in an increase in amplifier output current. The feedback principle employed makes the anemometer especially suited for measurement of high frequency flow fluctuations. Without feedback the upper frequency limit is restricted to approximately 100 Hz by the thermal capacity of the probe; with feedback the bandwidth increases by a factor approximately equal to the loop gain. Upper frequency limits of the amplifier are usually at least 100 kHz. The usual limitation is the probe which typically for $5\,\mu$ tungsten sensors has an upper frequency limit of 85 kHz.

The mean (D.C.) and R.M.S. (A.C.) components of the total instantaneous voltages existing across the probe are measured.

CALIBRATION OF HOT WIRES FOR VELOCITY CHANGE

Although it is possible to derive theoretical expressions for the necessary calibration constants, variations in wire composition, shape and tension make it necessary to individually calibrate each probe. Various expressions have been derived and used for calibration purposes, the usual expression being

$$E^2 = A + B(U)^n \tag{8.1}$$

where E is mean voltage; U is mean velocity; and n is a constant between 0·45 and 0·5.

This expression has been found to be unsatisfactory when used over a large cooling range. It has been found that higher order polynomials provide excellent curve fits to the experimental calibration curves of E vs U. The curve fit commonly used is of the form[15]

$$E^2 = A + BU^{1/2} + CU \qquad (8.2)$$

In calibration, a much better curve fit is obtained if the zero velocity voltage is excluded (i.e. $A = E_0^2$). A spurious value of A is generated owing to free convection effects. Typically, $A_{(U=0)} = E_0^2 = 8·52$ from measurement, while a value of 8·9 is obtained from the calibration curve.

Unfortunately, many hot wire probes have different calibration constants for the probe facing in different directions. The instantaneous cooling velocity is given by

$$U^2 = u^2 + G^2 w^2 + K^2 v^2 \qquad (8.3)$$

with values of $G = 1·15$ and $K = 0·25$ for miniature Disa probes. G and K must be determined for each individual probe as manufacturing variations affect these values. Several manufacturers are now producing probes in which $G \approx 1$ and $K = 0·05$. This is achieved by careful prong spacing and by gold plating the sensing wire near to the prongs.

FLOW VELOCITY MEASUREMENTS
For velocity measurements in flows where turbulence intensities are less than 20% and where one main flow direction predominates, the standard method for calculating turbulence intensity may be used[56]. The mean velocity is obtained directly from the mean voltage level, and the turbulence intensity from the R.M.S. voltage using the expressions

$$\bar{U} = \left(\frac{E^2 - A}{B}\right)^2 ; \qquad \frac{(U'^2)^{1/2}}{\bar{U}} = \frac{4(\overline{E'^2})^{1/2}\bar{E}}{\bar{E}^2 - \bar{E}_0^2}$$

A lineariser may be used between the anemometer, the voltmeter and the R.M.S. voltmeter.

In highly turbulent flows, the standard method of analysis of signals is not valid and linearisers cannot readily be used. Two methods of analysing hot wire anemometer signals have been developed.[15,16] In both methods six different measurements are taken at a given point to obtain

$$\bar{u}, \bar{v}, \bar{w}; \qquad \overline{u'^2}, \overline{v'^2}, \overline{w'^2}; \qquad \overline{u'w'}, \overline{v'w'}, \overline{u'v'}$$

Both methods give very similar results for a given set of input data. Two probes are used, at 45° and a straight probe. The instantaneous velocity vector at each probe position is calculated from the set of six measurements, *i.e.* expressions are derived for *u*, *v* and *w*. By assuming the six output voltages to be of similar regular shape and frequency distribution,

$$\bar{u}^2 + \overline{u'^2}, \qquad \bar{v}^2 + \overline{v'^2}, \qquad \bar{w}^2 + \overline{w'^2}$$

may be derived. If it is further assumed that the six output voltages are in phase, then $\bar{u}$, $\bar{v}$ and $\bar{w}$ may also be derived.

However, it has been found that these latter methods, whilst being an improvement on previous methods, still leave something to be desired when applied to recirculation zones and to the axis of swirling flows. Whilst axial and swirl velocities are much as expected, radial velocities are an order of magnitude too large in many positions. The swirl velocity is also too large near to the axis of symmetry, and measurements give a high swirl velocity on the axis of symmetry. The disadvantages of the six position methods may be summarised as:

(a) with six measurements having to be taken at a point, the flow state may alter (owing to the extended time needed for measurement), hence the calibration may alter over a number of readings;

(b) two probes are necessary which may not be placed in quite the same position;

(c) it is necessary to assume that all six voltages are regular, of similar frequency distribution and in phase.

The following method of analysis only requires one probe with four point measurements and has proved to be a much more accurate method of evaluating velocity levels in highly turbulent flows.

CALIBRATION OF HOT WIRE PROBES

Hot wire probes are usually calibrated in a uniform, low turbulence air stream. The air velocity is usually measured by a pitot tube or venturi meter. Figure 8.4 shows a hot wire probe with three components of a total velocity vector acting in directions parallel to the three axes. The three different calibration curves obtained are shown in Fig. 8.5. For directions *u* and *w* the calibration curves are very close, particularly for certain types of probe.[16] The difference is due to the wake effect of the prongs—the effect varies with the (L/d) ratio of the probe. In comparison,

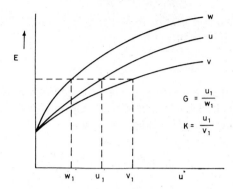

Fig. 8.5. Calibration curves for three different directions.

for direction v, the calibration curve is very different to the other two showing low sensitivity throughout the calibrated velocity range.

Curve fitting techniques are used to fit a curve of the form

$$E^2 = A + BU^{1/2} + CU \qquad (8.4)$$

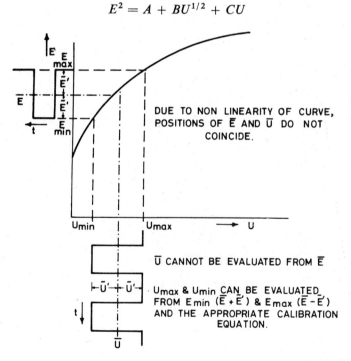

DUE TO NON LINEARITY OF CURVE, POSITIONS OF $\bar{E}$ AND $\bar{U}$ DO NOT COINCIDE.

$\bar{U}$ CANNOT BE EVALUATED FROM $\bar{E}$

U_{max} & U_{min} CAN BE EVALUATED FROM E_{min} $(\bar{E} + \bar{E}')$ & E_{max} $(\bar{E} - \bar{E}')$ AND THE APPROPRIATE CALIBRATION EQUATION.

Fig. 8.6. The effect of non-linearity on the respective positions of $\bar{E}$ and $\bar{U}$.

to the voltage velocity relationships for the three directions. Referring to Fig. 8.5, the response of the hot wire to the three velocity components is

$$U^2 = u^2 + G^2 w^2 + K^2 v^2 \qquad (8.5)$$

For direction u the calibration curve is used to evaluate U^2. The G and K functions, allowing for the direction properties of the probes, are evaluated from the respective calibration curves:

$$G = \frac{u}{w} \qquad K = \frac{u}{v} \, (E^2 \text{ constant})$$

G and K vary little with probe voltage—typical values for Disa gold line probes, $G \approx 1{\cdot}06$, $K \approx 0{\cdot}25$.

Owing to the non-linearity of the calibration curve, Fig. 8.5, assumptions must be made about the shape of the velocity signals to evaluate mean and fluctuating velocities. Consider the fluctuations are of square profile, Fig. 8.6, then the response of the hot wire is as shown.

It is apparent that the mean velocity and mean voltage points do not coincide on the calibration curve.

To evaluate velocity levels, the following procedure is adopted:

$$U_{\text{max}} = \left[\frac{-B + [B^2 - 4C(A - (\bar{E} + \tilde{E}')^2)]^{1/2}}{2C} \right]^2 = f(\bar{E} + \tilde{E}')$$
$$\qquad (8.6)$$

$$U_{\text{min}} = \left[\frac{-B + [B^2 - 4C(A - (\bar{E} - \tilde{E}')^2)]^{1/2}}{2C} \right]^2 = f(\bar{E} - \tilde{E}')$$
$$\qquad (8.7)$$

i.e. positions of minimum and maximum voltage correspond to positions of minimum and maximum velocity.

Hence

$$U^2_{\text{max}} = f(\bar{E} + \tilde{E}')^2 \qquad (8.8)$$

$$U^2_{\text{min}} = f(\bar{E} - \tilde{E}')^2 \qquad (8.9)$$

$$\bar{U} = \tfrac{1}{2}[f(\bar{E} - \tilde{E}') + f(\bar{E} - \tilde{E}')] \qquad (8.10)$$

$$\tilde{U}' = (\overline{U'^2})^{1/2} = \tfrac{1}{2}[f(\bar{E} + \tilde{E}') - f(\bar{E} - \tilde{E}')] \qquad (8.11)$$

Triangular and sine wave assumptions give similar results.

RESPONSE EQUATIONS

Consider the co-ordinate system shown in Fig. 8.7(a)—a straight hot wire probe is mounted parallel to the v-axis. The probe may be rotated

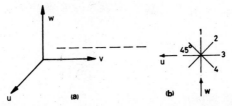

Fig. 8.7. *The four point measuring technique. (a) The co-ordinate system—probe parallel to v-axis; (b) each probe measuring system is separated by 45° of rotation.*

about its axis of symmetry and a four point measuring technique is used, Fig. 8.7(b).

Consider the instantaneous total velocity vectors acting on each of the four hot wire positions, Fig. 8.8:

$$U_1{}^2 = v^2 + G^2u^2 + K^2w^2 \tag{8.12}$$

$$U_2{}^2 = v^2 + \frac{G^2}{2}(w + u)^2 + \frac{K^2}{2}(w - u)^2 \tag{8.13}$$

$$U_3{}^2 = v^2 + G^2w^2 + K^2u^2 \tag{8.14}$$

$$U_4{}^2 = v^2 + \frac{G^2}{2}(w - u)^2 + \frac{K^2}{2}(w + u)^2 \tag{8.15}$$

We wish to solve these equations for u, w and v. Only three equations are required out of the four. The two sets of equations chosen are (8.12), (8.13) and (8.15) and (8.13), (8.14) and (8.15).

Take eqns. (8.13), (8.14) and 8.15):

$$U_2{}^2 + U_4{}^2 = 2v^2 + (G^2 + K^2)(w^2 + u^2) \tag{8.16}$$

$$U_2{}^2 - U_4{}^2 = (G^2 - K^?)\, 2wu \tag{8.17}$$

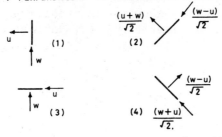

Fig. 8.8. *Velocity diagrams for various probe positions.*

$$U_2{}^2 + U_4{}^2 - 2U_3{}^2 = (G^2 - K^2)(u^2 - w^2) \tag{8.18}$$

From (8.17)

$$w = \left(\frac{U_2{}^2 - U_4{}^2}{2(G^2 - K^2)}\right)\frac{1}{u} \tag{8.19}$$

Substitute into (8.18) for w

$$U_2{}^2 + U_4{}^2 - 2U_3{}^2 = \left[u^2 - \left(\frac{U_2{}^2 - U_4{}^2}{2(G^2 - K^2)}\right)^2 \frac{1}{u^2}\right](G^2 - K^2) \tag{8.20}$$

Hence

$$u^4 - u^2[U_2{}^2 + U_4{}^2 - 2U_3{}^2]\frac{1}{(G^2 - K^2)} - \frac{1}{4}\left[\frac{U_2{}^2 - U_4{}^2}{(G^2 - K^2)}\right]^2 = 0 \tag{8.21}$$

and

$$u = \left(\left\{[(U_2{}^2 + U_4{}^2 - 2U_3{}^2) + ((U_2{}^2 + U_4{}^2 - 2U_3{}^2)^2 \right.\right.$$
$$\left.\left. + (U_2{}^2 - U_4{}^2)^2)^{1/2}]\frac{1}{2(G^2 - K^2)}\right\}\right)^{1/2} \tag{8.22}$$

A similar expression may be derived for w and then v may be derived by substituting back into (8.14) with $u^2 + w^2$.

Let

$$u = g(U_2{}^2, U_3{}^2, U_4{}^2) \tag{8.23}$$

Similarly, having derived expressions for w and v, let

$$w = h(U_2{}^2, U_3{}^2, U_4{}^2) \tag{8.24}$$

$$v = f(U_2{}^2, U_3{}^2, U_4{}^2) \tag{8.25}$$

EVALUATION OF THE EQUATIONS

To evaluate mean velocity and stress components from eqns. (8.23), (8.24) and (8.25) it is necessary to assume that three relevant voltages E_2, E_3 and E_4 are of similar shape and of similar frequency distribution (see Fig. 8.9a). It must also be assumed that E_2 and E_4 are in phase. To examine the validity of these assumptions it is necessary to re-examine the relevant equations, (8.12), (8.13), (8.14), (8.15).

If, say u is small, eqns. (8.13), (8.14) and (8.15) are of very similar form and hence likely to be of similar shape and frequency distribution. Equations (8.13) and (8.15) are virtually identical and the 'in phase' assumption seems to be reasonable.

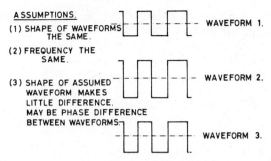

ASSUMPTIONS.
(1) SHAPE OF WAVEFORMS THE SAME. — WAVEFORM 1.
(2) FREQUENCY THE SAME.
(3) SHAPE OF ASSUMED WAVEFORM MAKES LITTLE DIFFERENCE. — WAVEFORM 2.
MAY BE PHASE DIFFERENCE BETWEEN WAVEFORMS — WAVEFORM 3.

Fig. 8.9(a). The wave form assumption.

In general some knowledge of the flow field is required beforehand, so that the probe may be aligned in the most suitable direction. Another benefit of such alignment is that the probe is most sensitive to the maximum velocity vector. For the respose equations (8.13), (8.14) and (8.15) the greatest sensitivity is to v and w components.

The phase difference between eqns. (8.14) and (8.13) and (8.15) may be derived as follows:

$$U_2{}^2 + U_4{}^2 = 2v^2 + (G^2 + K^2)(w^2 + u^2) = U_1{}^2 + U_3{}^2 \qquad (8.26)$$

Hence

$$U_1 = (U_2{}^2 + U_4{}^2 - U_3{}^2)^{1/2} \qquad (8.27)$$

Taking the time mean average of both sides of eqn. (8.27),

$$\bar{U}_1 = (\overline{U_2{}^2 + U_4{}^2 - U_3{}^2})^{1/2} \qquad (8.28)$$

N.B. The frequency and phase form of U_1 is irrelevant.

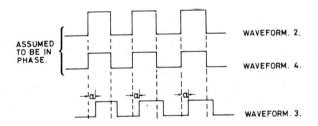

ASSUMED TO BE IN PHASE. — WAVEFORM. 2.
— WAVEFORM. 4.
— WAVEFORM. 3.

A PHASE SHIFT OF α RELATIVE TO WAVEFORM 2 & 4 IS ASSUMED — IT IS POSSIBLE TO EXPRESS THIS AS A FREQUENCY DIFFERENCE AS WELL.

Fig. 8.9(b). The phase shift assumption.

Let the phase difference between U_2, U_4 and U_3 be α_3 and assume a square waveform. Hence from Fig. 8.9(b)

$$2\bar{U}_1 = [(U_2{}^2{}_{max} + U_4{}^2{}_{max} - U_3{}^2{}_{max})^{1/2} + (U_2{}^2{}_{min} + U_4{}^2{}_{min} - U_3{}^2{}_{min})^{1/2}]$$

$$\times (1 - \alpha_3) + [(U_2{}^2{}_{max} + U_4{}^2{}_{max} - U_3{}^2{}_{min})^{1/2}$$

$$+ (U_2{}^2{}_{min} + U_4{}^2{}_{min} - U_3{}^2{}_{max})^{1/2}] \alpha_3 \qquad (8.29)$$

Hence

$$\alpha_3 = \frac{[2\bar{U}_1 - (U_2{}^2{}_{max} + U_4{}^2{}_{max} - U_3{}^2{}_{max})^{1/2} - (U_2{}^2{}_{min} + U_4{}^2{}_{min} - U_3{}^2{}_{min})^{1/2}]}{[(U_2{}^2{}_{max} + U_4{}^2{}_{max} - U_3{}^2{}_{min})^{1/2} + (U_2{}^2{}_{min} + U_4{}^2{}_{min} - U_3{}^2{}_{max})^{1/2} - (U_2{}^2{}_{max} + U_4{}^2{}_{max} - U_3{}^2{}_{max})^{1/2} - (U_2{}^2{}_{min} + U_4{}^2{}_{min} - U_3{}^2{}_{min})^{1/2}]} \qquad (8.30)$$

$$\bar{U}_1 = \tfrac{1}{2}(U_{1max} + U_{1min}) \qquad (8.31)$$

For the response equations (8.13), (8.14) and (8.15) the greatest sensitivity is to v and w components.

With the various assumptions made, mean velocities may be evaluated as follows:

$$\bar{v} = \frac{1}{T} \int_0^T f(U_2{}^2, U_3{}^2, U_4{}^2)\, dt \qquad \text{[from (8.13)]} \qquad (8.32)$$

Assume a regular waveform (say square) for $U_2{}^2$, $U_3{}^2$ and $U_4{}^2$; $U_2{}^2$ and $U_4{}^2$ are in phase.

With the phase difference α_3 between U_3 and U_2 from eqn. (8.29)

$$\bar{v} = \tfrac{1}{2}\{[f(U_2{}^2{}_{max}, U_3{}^2{}_{max}, U_4{}^2{}_{max}) + f(U_2{}^2{}_{min}, U_3{}^2{}_{min}, U_4{}^2{}_{min})]$$

$$\times (1 - \alpha_3) + [f(U_2{}^2{}_{max}, U_3{}^2{}_{min}, U_4{}^2{}_{max})$$

$$+ f(U_2{}^2{}_{min}, U_3{}^2{}_{max}, U_4{}^2{}_{min})]\alpha_3\} \qquad (8.33)$$

$f(U_2{}^2{}_{max}, U_3{}^2{}_{max}, U_4{}^2{}_{max})$ may be evaluated by substituting their values into eqn. (8.25) for $U_2{}^2$, $U_3{}^2$, $U_4{}^2$ respectively. The other terms in eqn. (8.33) may be evaluated in a similar manner and hence $\bar{v}$ may be evaluated. Similar results apply for $\bar{v}$ and $\bar{w}$.

Evaluation of turbulent fluctuating velocities
Now

$$\overline{v^2} = \overline{(v + v')^2} = \bar{v}^2 + \overline{v'^2} \qquad (\textit{see Hinze}[17]) \qquad (8.34)$$

From eqn. (8.25)

$$v^2 = [f(U_2{}^2, U_3{}^2, U_4{}^2)]^2$$

Hence

$$\overline{v'^2} = \overline{f(U_2{}^2, U_3{}^2, U_4{}^2)^2} - \bar{v}^2 \tag{8.35}$$

$\overline{f(U_2{}^2, U_3{}^2, U_4{}^2)^2}$ may be evaluated in the same way as $\overline{f(U_2{}^2, U_3{}^2, U_4{}^2)}$:

$$\begin{aligned}\overline{v'^2} = \{&[f(U_2{}^2{}_{max}, U_3{}^2{}_{max}, U_4{}^2{}_{max})^2 \\
&+ f(U_2{}^2{}_{min}, U_3{}^2{}_{min}, U_4{}^2{}_{min})^2](1 - \alpha_3) \\
&+ [f(U_2{}^2{}_{max}, U_3{}^2{}_{min}, U_4{}^2{}_{max})^2 \\
&+ f(U_2{}^2{}_{min}, U_3{}^2{}_{max}, U_4{}^2{}_{min})^2]\alpha_3\}/2 - \bar{v}^2\end{aligned} \tag{8.36}$$

Similar results apply for $\overline{u'^2}$ and $\overline{w'^2}$.

Evaluation of shear stresses

$$\overline{wv} = \overline{(w + w')(u + u')} = \bar{u} \times \bar{w} + \overline{u'w'} \quad (see \text{ Hinze}[17]) \tag{8.37}$$

Hence

$$\overline{u'w'} = \overline{wu} - \bar{u} \times \bar{w} \tag{8.38}$$

Now

$$\overline{wu} = \overline{h(U_2{}^2, U_3{}^2, U_4{}^2) \times g(U_2{}^2, U_3{}^2, U_4{}^2)} \tag{8.39}$$

Hence

$$\begin{aligned}\overline{u'w'} = \tfrac{1}{2}\{&[h(U_2{}^2{}_{max}, U_3{}^2{}_{max}, U_4{}^2{}_{max}) \times g(U_2{}^2{}_{max}, U_3{}^2{}_{max}, U_4{}^2{}_{max}) \\
&+ h(U_2{}^2{}_{min}, U_3{}^2{}_{min}, U_4{}^2{}_{min}) \times g(U_2{}^2{}_{min}, U_3{}^2{}_{min}, U_4{}^2{}_{min})](1 - \alpha_3) \\
&+ [h(U_2{}^2{}_{max}, U_3{}^2{}_{min}, U_4{}^2{}_{max}) \times g(U_2{}^2{}_{max}, U_3{}^2{}_{min}, U_4{}^2{}_{max}) \\
&+ h(U_2{}^2{}_{min}, U_3{}^2{}_{max}, U_4{}^2{}_{min}) \\
&\qquad \times g(U_2{}^2{}_{min}, U_3{}^2{}_{max} U_4{}^2{}_{min})]\alpha_3\}/2 - \bar{w} \times \bar{u}\end{aligned} \tag{8.40}$$

Similar results apply for $\overline{v'w'}$ and $\overline{u'v'}$.

The waveform assumption

The assumption of a square waveform for $U_2{}^2$, $U_2{}^3$ and $U_4{}^2$ is purely arbitrary, a triangular or sine wave being just as applicable. The results obtained with different shaped waveforms are very similar; the square waveform, however, is the most convenient to use.

The second set of equations

As mentioned previously, the equations may be solved in a different order for the mean and fluctuating velocity components. A similar set of expressions for u, w and v is obtained except eqns. (8.12) (8.13) and (8.15) are used. The phase difference α_1 between U_2, U_4 and U_1 may be determined in the same manner as α_3. Hence $\bar{u}$, $\bar{v}$, $\bar{w}$, $\overline{u'^2}$, $\overline{v'^2}$, $\overline{w'^2}$, $\overline{u'w'}$, $\overline{v'w'}$, $\overline{u'v'}$ may be evaluated as before. The choice of results depends to a large extent on the probe positioning and flow field. As an example, suppose v and w are large and u is small. Then eqns. (8.13), (8.14) and (8.15) are those most likely to be of similar form and frequency, are those most sensitive to v and w and so should be used to evaluate $\bar{v}$, $\bar{w}$, $\overline{v'^2}$, $\overline{w'^2}$, $\overline{v'w'}$. It should be noted that these equations are relatively insensitive to u. Equations (8.12), (8.13) and (8.15) are much more sensitive to u and should be used for evaluation of $\bar{u}$, $\overline{u'^2}$, $\overline{u'w'}$ and $\overline{u'v'}$.

Velocity direction

The hot wire probe is sensitive to velocity direction when orientated in directions 2 and 4.

From eqn. (8.19)

$$wu = \frac{U_2{}^2 - U_4{}^2}{2(G^2 - K^2)}$$

Hence taking time mean averages of both sides of the equation,

$$\overline{wu} = \frac{\overline{U_2{}^2} - \overline{U_4{}^2}}{2(G^2 - K^2)} = \overline{w'u'} + \bar{w} \times \bar{u} \qquad (8.41)$$

Hence

$$\bar{w} \times \bar{u} = \overline{wu} - \overline{w'u'} \qquad (8.42)$$

[In eqn. (8.39) $\overline{wu}$ is always positive, but in eqn. (8.41) $\overline{wu}$ may be negative.]

Providing one velocity direction is known the other velocity direction may be determined from the sign of eqn. (8.42).

Limitations

The voltage/velocity relationship of the hot wire probe is of the form

$$E^2 = A + BU^{1/2} + CU$$

The probe is therefore not sensitive to local flow reversals because the voltage response is of E^2 form. The response of the hot wire becomes

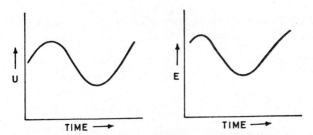

(a) LOW TURBULENCE INTENSITY - OUTPUT VOLTAGE
FOLLOWS U.

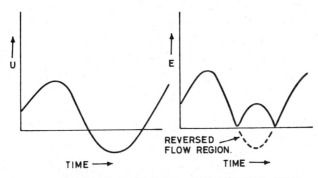

REVERSED
FLOW REGION.

(b) HIGH TURBULENCE INTENSITY – INSTANTANEOUS FLOW
REVERSAL – THE OUTPUT VOLTAGE CANNOT FOLLOW U–
SPURIOUS MEAN AND FLUCTUATING VOLTAGES
GENERATED.

Fig. 8.10.   Limitations of the method.

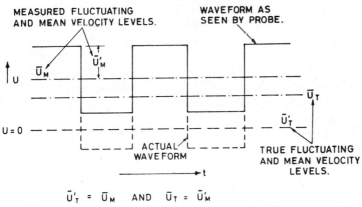

MEASURED FLUCTUATING
AND MEAN VELOCITY LEVELS.

WAVEFORM AS
SEEN BY PROBE.

TRUE FLUCTUATING
AND MEAN VELOCITY
LEVELS.

$$\bar{U}'_T = \bar{U}_M \quad \text{AND} \quad \bar{U}_T = \bar{U}'_M$$

Fig. 8.11.   The effect of instantaneous flow reversal on the square wave.

spurious as the instantaneous flow reversal is interpreted as a positive velocity (see Fig. 8.10). The limits may be summarised as follows:

Shape of assumed waveform	Onset of instantaneous reverse flow
Square wave	100 % turbulence intensity
Sinusoidal wave	71 % Turbulence intensity
Triangular wave	58 % Turbulence intensity

Corrections can be applied for instantaneous flow reversal (Fig. 8.11), but they rely on a knowledge of its occurrence which is difficult to acquire.

8.3 PRESSURE

STATIC PRESSURE—DISC PROBE

An accurate measurement of the static pressure in turbulent shear flow is difficult owing to the influence of the velocity and pressure fluctuations on the pressure measuring device which is introduced into the flow. For the study of swirl and flow in the wake of stabiliser discs where there are significant radial and axial pressure gradients, the measurement of the static pressure distributions is required in order to determine the flow pattern. A review of methods for measuring static pressure has been made by Bryer et al.[18] Ricou and Curtet[19] and Miller and Comings[20] have shown the reliability of the results obtained with a disc probe in both single and dual jet flows. By measuring both the mean and fluctuating components of velocity with a hot wire anemometer, they could obtain a balance between the pressure and shear terms in the equations of motion. A slight correction term is still required to take into account the intensity of turbulence, but the measurements are more accurate than those obtained with the static holes of a standard pitot tube. In strongly swirling flows there may be a significant change in the static pressure between the total head and static holes of a normal pitot and it was found necessary to measure the total head with an impact-probe and subsequently the static pressure with a disc probe at the same position.

The disc probe shown in Fig. 8.12 consists of a thin metal disc 20 mm diameter with a central static pressure measuring hole of 1 mm. On the underside of the disc two tubes are fitted with their edges sawn at 45°. In order to measure the static pressure it is necessary to align the disc so that the resultant velocity is in the plane of the disc at each measuring point. For flows in which there are only radial and axial components of velocity it is unnecessary to rotate the probe.

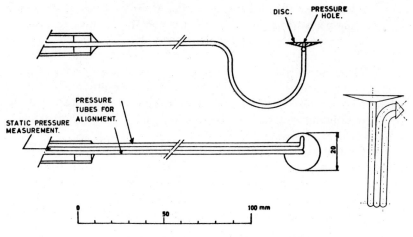

Fig. 8.12. *Static pressure probe (uncooled).*

However, when tangential components of velocity are present, it is necessary to rotate the probe until the differential pressure measured by the two side tubes is zero. The static pressure is then measured directly from the central hole with the aid of a pressure transducer.

8.4   TEMPERATURE

FINE WIRE THERMOCOUPLES

The principal advantages of using a fine wire thermocouple for temperature measurements are (*i*) measurements can be made with a high precision characteristic of electrical measurements; (ii) extremely small thermocouples can be constructed so that a high resolution can be obtained and the disturbance to the flame minimised; and (iii) the method is simple and convenient and measurements can be taken without the use of costly and complex equipment. There are, however, a number of sources of error which should be taken into account.

In general, the probe does not obtain the same temperature as the hot gas. There is a radiative energy exchange between the flame surroundings and the probe. The probe usually radiates energy to cooler walls and, in the steady state, this is balanced by convective energy transfer from the gas to the probe. This means that the wire is at a lower temperature than the gas, the value of this temperature difference being known as the radiation loss. Since the radiation from the probe increases as the fourth power of the probe temperature, the correction which must be made for this loss increases very sharply with temperature. Consideration of the energy balance for the probe also shows that the magnitude of the correction decreases with the size of the probe, *e.g.* at a gas temperature of 1300°K a 0·0005 in (12 μm) diameter platinum wire probe may have a radiation correction of 5°K rising to 60°K for a gas temperature of 2100°K. The corresponding values of the correction for a 0·002 in (50 μm) diameter probe would be 35°K and 200°K.

The disadvantages of very fine wires lies in their fragility. The wires need to be supported by thicker wires of the same material and energy is conducted along the fine wires to the supports. This conduction can cool the hot junction and the amount by which it does so depends upon the length of wire between the junction and the support. The decrease in hot junction temperature as a result of this effect is called the conduction loss.

There is also a need to take into account the complications of catalytic heating on the wire surface. At temperatures up to 1500°C fine wires may be protected by a very thin coating of silica, which can be readily deposited on the wire surface. The necessity for a completely continuous coating is emphasised. It is not possible to coat successfully on untidy or badly formed junctions with protruding wire ends.

Account can also be taken of the effect of the stream velocity Reynolds number. As the Reynolds number increases, the conduction cooling error decreases.

Calculation of all these corrections can only be approximate and the complete method is given in reference 21.

Dixon-Lewis and Isles[22] also investigated the effect of thermocouple configuration on the measured profile across a flat flame. They found that the optimum configuration was one in which the immediate leads to the hot junction lay in an isothermal plane in the flame.

Smith and Gordon[23] used simple hairpin shaped thermocouples in their investigation of a methane diffusion flame. Two sizes of thermocouple were constructed, one from 25 μm wire and the other from 75 μm wire. Both types were used bare and coated with silica. Inside the luminous zone of the flame, where there were extremely low concentrations of oxygen and low temperature gradients, essentially the same temperatures were recorded by both sizes of thermocouple whether covered or bare. Towards the edge of the flame where the oxygen concentrations rapidly increase, the 25 μm and 75 μm bare wire thermocouples showed a significantly higher temperature than the coated thermocouples, indicating that the oxidation reactions were being catalysed. The 25 μm coated thermocouple was used as being the most accurate. They also attempted to correct for radiation and conduction losses and concluded that, in their case, the radiation losses were unimportant.

SUCTION PYROMETER

When a bare thermocouple is introduced into a flame, errors arise owing to the radiation exchange between the thermocouple and its surroundings. In the suction pyrometer, the thermocouple is shielded from its surroundings and gas is drawn over the sheathed thermocouple with a high suction velocity. Refractory shields minimise the gain or loss of radiant heat by a sheathed thermocouple and the high gas velocity makes convection the prevailing form of heat transfer from the gas to the thermocouple. The temperature measured will be the local gas temperature if it can be assumed that the presence of the refractory shield has no effect upon the local rate of combustion. The suction pyrometer, Fig. 8.13, consists of a platinum–platinum/10% rhodium thermocouple, protected by a sintered alumina sheath surrounded by two concentric silimanite radiation shields. The gases are drawn between the radiation shields and also between the inner shield and the thermocouple sheath with high velocity ($U \geq 40$ m/sec).

A number of different forms of the outer shield have been tried, and it was found experimentally that best results are obtained when the end of the shield is closed and two holes are made in the side walls of the outer shield as shown in Fig. 8.13.[7,8] One hole faces the burner and particles of dust and soot pass directly through the shield, thereby considerably reducing the amount of blockage by the particles.

The best position of the thermocouple element in relation to the entrance hole was experimentally determined and found to be 4 cm downstream from the edge of the hole.

The accuracy of the double shield suction pyrometer is dependent upon the suction velocity, the intensity of radiation and the geometry of the system. If the suction velocity is steadily increased from zero to the maximum obtainable with the system and a graph is plotted of the temperature measured by the thermocouple versus suction velocity, the position of the asymptote on the temperature axis will be the true gas temperature. A measure of the efficiency of the instrument can be obtained at any point in the furnace by measuring the temperature without suction and then with the maximum suction velocity:

$$\text{efficiency} = \frac{T_{(\text{max. suction velocity})} - T_{(\text{without suction})}}{T_{(\text{true gas temperature})} - T_{(\text{without suction})}}$$

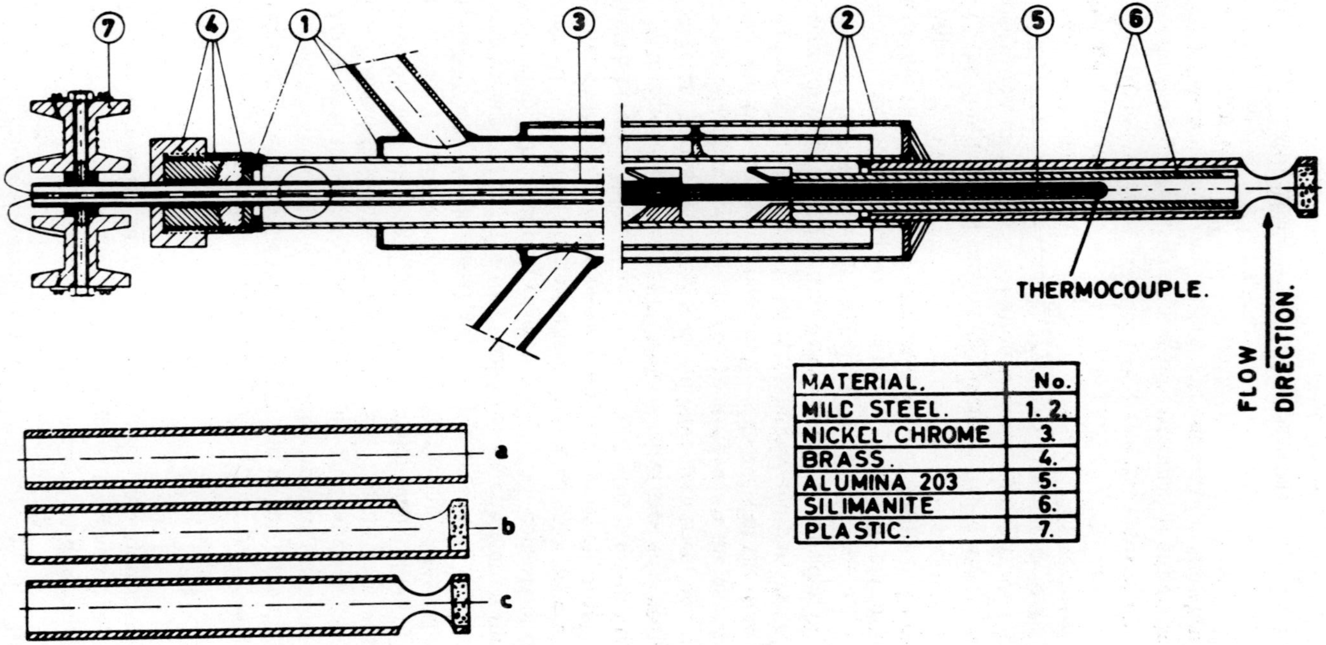
7
4
1
3
2
5
6
THERMOCOUPLE.
FLOW
DIRECTION.
a
b
c

MATERIAL.	No.
MILD STEEL.	1. 2.
NICKEL CHROME	3.
BRASS.	4.
ALUMINA 203	5.
SILIMANITE	6.
PLASTIC.	7.

Fig. 8.13. *Suction pyrometer.*

the 'true gas temperature' can also be determined by the use of the sodium line reversal method.[3] A direct comparison of temperatures measured by this spectroradiometric method and by a suction pyrometer showed an error of $\pm 6°C$ at $1600°C$ in the suction pyrometer measurement.[8]

Care has to be taken when measurements are made in the fuel-rich and high velocity regions of flames because the refractory shield may act as a bluff body stabiliser and thus lead to considerable errors in the measurement.

VENTURI PNEUMATIC PYROMETER

This method of temperature measurement developed by BCURA[24] is based on the fact that the gas density is inversely proportional to the absolute temperature. The instrument consists essentially of two venturi throats in series through which the gas drawn from the furnace has to pass (Fig. 8.14). One of the venturi throats is right at the end of the probe, where the hot gases enter. The second venturi is farther downstream so that the gases can cool down to an easily measurable temperature before reaching it.

From the conservation of mass of the gas passing through both venturies it follows that:

$$K_1 p_1 T_2 = K_2 p_2 T_1 \qquad (8.43)$$

and hence

$$T_1 = \frac{K_1 p_1}{K_2 p_2} T_2 \qquad (8.44)$$

where K_1 and K_2 are constants of the venturi throats, p_1 and p_2 are the pressure differences at the venturi throats and T_1 and T_2 are the absolute temperatures of the gas in the two venturies. The constants K_1 and K_2 are determined from a calibration with cold air. p_1, p_2 and T_2 are measured directly. T_1 can then be calculated.

Beaudouin manometers are used for the measurement of the pressure differences. These give an electric output proportional to the pressure difference. The output of these manometers together with that from the gas temperature measurement (T_2) can conveniently be fed to an analogue computer and the output of the computer can then be recorded directly in degrees of temperature.

The advantages of the venturi pneumatic pyrometer are that it enables measurements to be made in very high temperature regions ($1700°C$), and also that it can record instantaneous temperature fluctuations of the flame. The temperatures so measured are local gas temperatures, as in the

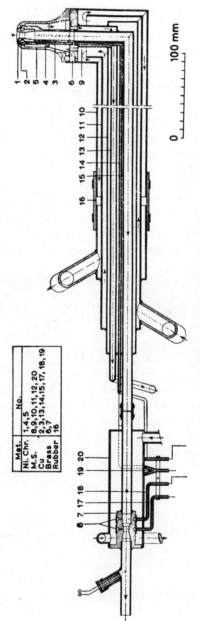

Mat.	No.
Ni. Chr.	1,4,5
M.S.	8,9,10,11,12, 20
Cu	2,3,13,14,15,17,18,19
Brass	6,7
Rubber	16

100 mm

0

Fig. 8.14. Venturi pneumatic pyrometer.

case of the suction pyrometer. A comparison of suction pyrometer readings with those of a venturi pneumatic pyrometer at Ijmuiden showed an agreement within $\pm 15°C$ at flame temperatures (1500°–1600°C). In regions of high fuel concentrations, however, the agreement was not as good. Temperatures measured by the venturi pyrometer were generally lower by 50 to 150°C than those measured by the suction pyrometer.

The instrument can be used in heavily dust laden gases, and for measurements in combustion chambers where there are high levels of radiation intensity it is not necessary to correct for radiation shield errors as in the case of the conventional suction pyrometer. The time response of the instrument is considerably shorter than that of the suction pyrometer and is about 3 sec.

The analogue computer uses two transducers to convert the two differential pressure signals into electrical signals, one of which is modified by the resistance thermometer signal giving the temperature at the cold end venturi. Two signals are then presented to a self balancing ratio recorder the indication of which is proportional to temperature at the hot end. The constant K is conveniently compensated for by setting the system to read room temperature when air is sucked through the probe before use in the furnace.

RESISTANCE THERMOMETERS

A constant temperature hot wire anemometer with a suitable probe may be used for mean temperature measurements providing the local velocity components are known.

The heat balance for a given hot wire is given by

$$P = (A + BU^{1/2} + CU)(T_s - T_e) \qquad (8.45)$$

where P is the power dissipated and A, B, C are calibration constants which vary with temperature but which can be allowed for.

For low turbulence levels ($< 20\%$)

$$\bar{P} = (A + B\bar{U}^{1/2} + C\bar{U})(T_s - T_e)$$

Knowing $\bar{U}$, $\bar{P}$ and T_s, the environmental temperature, T_e, may be calculated.

The method may be extended to obtain fluctuating temperature values. The limitation is that the heat balance equation depends upon the temperature difference between the hot wire and the environment. As this decreases, the sensitivity of the probe decreases. The maximum temperature for Pt hot wires (Disa) is about 1600°C which means that only temperatures up to about 1000°C may be measured. A cooled probe eliminates many of these difficulties. Such a probe is available commercially from Thermosystems (USA) Inc. The hot wire probe is sensitive to concentration fluctuations. Fortunately, in premixed gas–air systems these tend to be small due to the 80% N_2 present.

8.5   HEAT TRANSFER

Conductivity Plug-type Heat Flow Meters

Temperature gradients are measured in a plug of known thermal conductivity, the front face of which is heated and the rear face water-cooled. This enables the heat flux to the face of the plug to be calculated.

The instrument consists of a central plug usually made of copper or stainless steel surrounded by two guard rings of the same material as the plug (Fig. 8.15). The front faces of the guard rings are flush with that of the plug and their ends are water-cooled. In this way the guard rings prevent heat transfer from the central plug in the radial direction. Inside the central plug there are two Ni–NiCr thermocouples, separated by a known

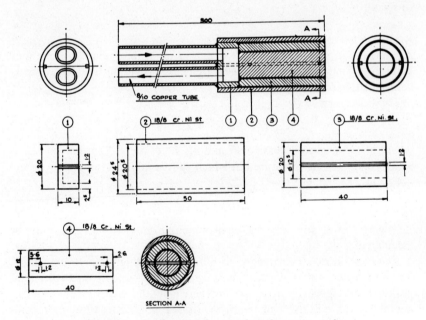

Fig. 8.15.   Conductivity plug-type heat flow meter probe.

distance along the axis of the plug, for the measurement of the temperature gradient.

When calibrating the heat flow meter in a 'black body' furnace we can write for the heat balance of the plug

$$\varepsilon_s \sigma T_B{}^4 = \frac{k}{s} \Delta T + \varepsilon_s \sigma T_s{}^4 \tag{8.46}$$

where:

k is the thermal conductivity of the plug;
s is the distance between the thermocouples;
T_B is the temperature of the black body furnace;
T_s is the plug surface temperature;
ΔT is the temperature difference as indicated by the thermocouples;
ε_s is the emissivity of the plug surface;
σ is the Stefan–Boltzmann constant.

The value of T_s is found by extrapolation assuming that the temperature gradient in the plug is constant. Results of an experiment in which the plug surface was covered with a thin soot layer so that its emissivity was approximately equal to unity have shown that the extrapolation of the plug temperature to the surface is a good approximation.[8]

Under experimental conditions when the heat flow meter is used in the furnace it is not practicable to cover the plug surface with soot because the soot layer may burn away in an oxidising atmosphere—and this can result in a change in surface emissivity during the measurement. The stainless steel plug, however, soon develops an oxide layer on its surface, which tends to maintain the surface emissivity constant.

The response time of the plug-type heat flow meter is relatively long; it could be as much as 5 min for the case when the plug surface temperature T_s is 970°K, and the heat flux to the plug surface is 8 cal/cm^2 sec. The emissivity of the oxidised surface of the stainless steel plug varied within the range

$$\varepsilon_s = 0.89 \text{ to } 0.93$$

Rates of total heat flow from the flame to points on the hearth of the Ijmuiden furnace were measured by a stainless steel conductivity-type heat flow meter, mounted on a probe with the plug axis perpendicular to the axis of the probe in order to facilitate the traversing of the furnace hearth (Fig. 8.16).[25]

A hollow ellipsoidal radiometer was used to measure the hemispherical radiation from the flame and the furnace walls to points on the hearth.

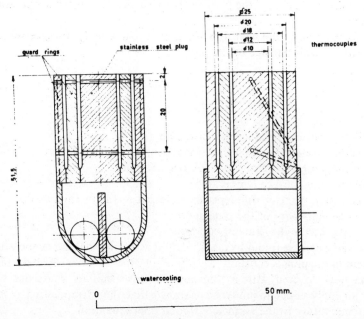

Fig. 8.16.   *Conductivity plug-type heat flow meter for measurements of total heat flow to the hearth of a furnace.*

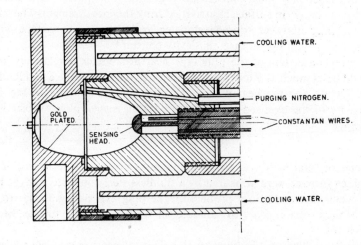

Fig. 8.17.   *Hollow ellipsoidal pyrometer for the measurement of hemispherical radiation to a surface element in the furnace.*

Figure 8.17 shows the arrangement of the radiometer as used for hearth measurements.[25]

The total heat flow to the conductivity plug heat flow meter can be written as

$$Q_t = Q_c + Q_r = h(T_g - T_s) + \varepsilon_s\sigma(T_R{}^4 - T_s{}^4) \qquad (8.47)$$

In the above equations $\sigma T_R{}^4$ represents the hemispherical radiation of the flame and of the furnace walls in terms of black body radiation as measured by the hollow ellipsoidal radiometer. The surface temperature of the plug, T_s, is determined by linear extrapolation of the measured plug temperature distribution to the surface and the surface emissivity ε_s is determined by calibration. The gas temperature adjacent to the plug surface, T_g, has to be measured separately.

Another method to differentiate between radiant and convective proportions of heat transfer to a point is by using two heat flow meters with different surface temperatures (*e.g.* calorimeter-type and plug-type heat flow meters or copper and stainless steel plug-type heat flow meters).

We can write for the heat flow meter with the high temperature surface

$$Q_1 = \varepsilon_1\sigma(\varepsilon_F T_F{}^4 - T_{s1}{}^4) + h(T_g - T_{s1}) \qquad (8.48)$$

and for the heat flow rate to the low surface temperature heat flow meter

$$Q_2 = \varepsilon_2\sigma(\varepsilon_F T_F{}^4 - T_{s2}{}^4) + h(T_g - T_{s2}) \qquad (8.49)$$

where:

Q_1 =heat flow rate to the high surface temperature heat flow meter;

Q_2 =heat flow rate to the low surface temperature heat flow meter;

h =coefficient of convective heat transfer;

T_g =temperature of gas adjacent to the heat flow meter;

T_F =mean radiant furnace temperature;

T_{s1} and T_{s2}=surface temperatures of the heat flow meters;

ε_1 and ε_2 =surface emissivities of the heat flow meters.

When $\varepsilon_1 = \varepsilon_2 = \varepsilon_s$ we can write for the coefficient of convective heat transfer

$$h = \frac{Q_2 - Q_1 - \varepsilon_s\sigma(T_{s1}{}^4 - T_{s2}{}^4)}{T_{s1} - T_{s2}} \qquad (8.50)$$

Under conditions where the convective proportion of the heat flow is small, the radiometer–heat flow meter method is preferable to the two heat flow meter method because the latter involves a larger number of calibrations and is therefore less accurate. In particular when the surface emissivities differ, *i.e.* $\varepsilon_1 \neq \varepsilon_2$, the radiation from the furnace walls and the flame, $\varepsilon_F \sigma T_F^4$, has to be measured separately and then the two heat flow meter method has lost its appeal of greater simplicity.

8.6 GAS AND SOLID PARTICLE CONCENTRATION MEASUREMENTS

The sampling of gas and solid particles from flames presents two main problems:

1. the sampling has to be 'iso-kinetic' in order to minimise the interference of the sampling probe with the flow pattern upstream of the sampling nozzle;
2. it is necessary to quench the gases in order to prevent any further reaction inside the sampling probe.

THE SAMPLING PROBE (FIG. 8.18)
The probe consists of a water jacket specially shaped to protect the sampling head and filter. Inside the separately water-cooled, demountable head is a large metal block to direct the stream of water along the walls. The sliding fit sampling tube is thus well cooled and provides efficient quenching. The probe and interior section tubes are made of stainless steel. The removable sampling tube is 50 mm long so that, for each of the various diameters employed, the temperature at the filter chamber is lower than 300°C. The tube diameter is chosen at each sampling point so that a sufficient volume of iso-kinetic sample may be withdrawn within a reasonable time for the given velocity.

A sintered bronze filter with a maximum passage diameter of 10 μm and a mean pore size of 5 μm has been found suitable. The filter elements have a surface area of 20 cm^2 and are about 2 mm thick.

Solids are collected in the filter and water vapour is condensed in a condenser after which the dry gases pass through a valve controlling the flow through the system.

THE ACCURACY OF SOLID CONCENTRATION MEASUREMENT. QUENCHING
Measuring errors may arise due to volume measurements of the sampled gas or in the weighing of particles collected in the filter. The factors affecting

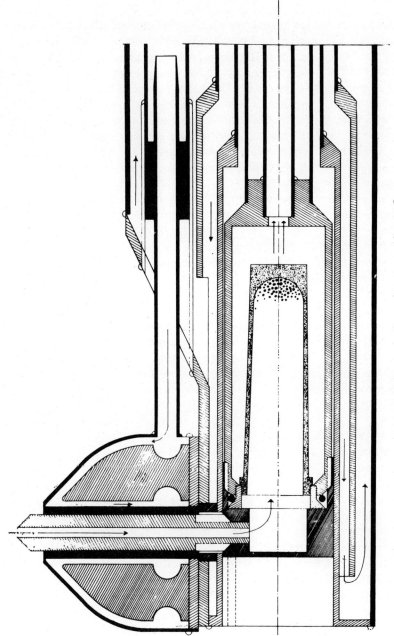

Fig. 8.18. *Sampling probe for iso-kinetic sampling in flames.*

the former are the gas meter error which in full flow is 1/1000 and its error in stopping and starting (± 20 cm^3). The dead volume of the probe and pump are, respectively, zero (because of effective purging) and 36 cm^3. The humidity of the atmosphere also introduces an error of between 0·5 % and zero. The sum total of errors is estimated as being less than 1·5 % in solids and between 0–50 mg litre^{-1} in the gas volume.

The accuracy of obtaining carbon weight is related to the amount sampled, the amount passing the filter and the condition of the sample circuit. The total weight including nozzle, carbon and filter is about 100 g weighed to 1/10 mg, whereas the carbon sample weight may vary from 100 mg to a few grammes.

A series of tests were carried out in the Ijmuiden furnace[8] in order to evaluate the errors and investigate the reproducibility of results in oil flames.

It was concluded that sampled carbon tends to form its own filter by depositing inside the pores of the bronze filter, thus the initial filter pore size is not very important.

The results showed that a negligible weight of carbon passes through the sintered bronze filter.

Measurements have been made of the gas temperature along the path of the gas in the sampling probe while taking samples in an oil flame. It was shown that the important variable is the rate of sampling (litres per minute) and that the speed at which the sample was cooled was inversely proportional to the rate of flow of the sampled gas.

CONCENTRATION MEASUREMENTS IN GAS FLAMES
The most successful technique that has been applied to date for the determination of the concentrations of stable species in flame systems is direct probe sampling. In principle this is a simple procedure; a suitable probe is inserted into the flame and a sample is withdrawn, quenched and analysed. This method only allows for the determination of species which are stable to the sampling process. This, however, includes all the quantitatively important species.

A sampling probe must be designed to produce a minimum disturbance to the flame both physically and compositionally. It must therefore be as small as possible and also the rate at which the sample is withdrawn must be such that the composition at the sampling point is not changed. Until recent years, water-cooled probes sampling at stream velocity (iso-kinetic sampling) were the only possibility, although they were known to be unsatisfactory in many respects. Fristrom with others[2] developed a completely new sampling technique which has proved to be a considerable advance. The principle of this technique is the use of a sonic-orifice. The pressure drop through the orifice is sufficient to quench the reactions and hence no cooling of the probe is necessary. The probes are constructed from fine quartz tubing with one end tapered down to a fine orifice. Considerable evidence is now available that this technique is by far the best available. It has been shown that the results obtained using such a probe are independent of the sampling pressure, the size of sampling orifice, the orientation of the probe, the precise construction of the probe and whether it is cooled or not. When tested for sampling bias in a known concentration gradient, none was found, and a theoretical analysis treating the probe as a sink showed that the compositional and physical disturbances of a flame by the probe were small.

8.7 MEASUREMENT OF DROPLET SIZES, VELOCITIES AND ANGLE OF FLIGHT USING HIGH SPEED PHOTOGRAPHY

Research is being carried out at Sheffield on the aerodynamics of the interaction between both isothermal and burning liquid sprays with air flows.

The measurements required from these sprays are droplet size, velocity, angle of flight and concentration. A survey of drop size measurement techniques showed that the only method which was suitable was high speed photography, either using cine film cameras, high speed shutters or short duration light sources. Of these, the latter appeared to be the most suitable and spark sources were chosen for the work.

The technique which is described below has been successfully applied to the photography of an isothermal pressure jet spray in a uniform air flow field,[28] and is presently being extended to cover the photography of twin fluid atomiser flames and also both isothermal and burning oil sprays stabilised by bluff bodies (see Section 6.4). The experimental system is shown in Fig. 8.19.

Essentially the sequence of events was to open the camera shutter, fire the first spark, and then, after a short and variable time interval, to fire the second spark. The camera shutter was then closed. This procedure resulted in a photographic record of the droplets in a given position within the spray at two points in time separated by a short interval. It is important to realise that only a two-dimensional section of the spray is photographed. Suitable projection and magnification of the photographic plate on to a screen made possible the measurement of distances, angles of flight, diameters and concentrations of droplets. Measurement of the time interval between the sparks with an optical probe unit coupled to an oscilloscope, the display of which was photographed with a polaroid camera, made possible the calculation of velocity.

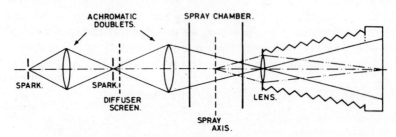

Fig. 8.19.   *Optical system for double image photography.*

Photographs of the droplets taken in the above manner were obtained at various heights in the two-dimensional plane of the spray by traversing the atomising unit vertically and the optical system, including the camera, in a horizontal plane.

It was necessary for the sequence of events outlined above to occur so quickly that a system of electronic control was introduced. The measurements of distance, velocity, angle and concentration were made directly from the projected photographic plates with the aid of a data logging system, the output from which was in a form suitable for feeding directly into the computer.

Individual parts of the apparatus are now described in more detail.

THE SPARK UNITS
A diagram of the spark units is shown in Fig. 8.20.

The electrode holders were constructed from $2\frac{1}{2}$ in brass tubing with a $1\frac{1}{2}$ in diameter hole bored through them to allow the uninhibited passage of light. They were placed on an optical bench, the support fitting into a standard saddle. The brass tube and electrodes were horizontal and inclined at 90° to the optical bench.

The electrodes were constructed from $\frac{1}{4}$ in od copper tubing with rounded inserts placed in the ends. (Although Fig. 8.20 shows three separate electrodes, the trigger electrode was placed inside the earth

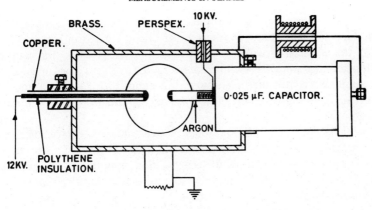

Fig. 8.20.   Electrode holder (double image photograph).

electrode.) The trigger electrode consisted of uniradio 43 coax cable with all but the polythene coating removed. This proved to be adequate insulation between the two electrodes. When a trigger pulse was passed a spark occurred between the central copper wire of the coax cable and the outer earthed copper tubing. This spark caused the air in the spark gap to be ionised and the capacitor charged to 10 kV discharged to the earth electrode. Both electrodes were adjustable and provision was made on the positive electrode for the introduction of argon to overcome humidity variations. This was, however, found to be unnecessary.

PHOTOGRAPHY AND OPTICAL SYSTEM
Capacitor discharge in air emits light over a range of wavelengths but mainly in the blue and near ultraviolet regions of the spectrum. Tests also showed, however, that some green and red light is present.

For isothermal work monochromatic plates are used (Kodak B.10) and, with $\frac{1}{2}$-sec exposure times, the photography can take place in normal daylight conditions although neon lighting should be extinguished. Processing under a red safety light is possible.

More recent results obtained using a flame produced by a twin fluid atomiser indicate that either orthochromatic or panchromatic film is more sensitive and provides greater contrast than the B.10 plates but shorter exposure times are required. In addition, the panchromatic films must be processed in total darkness. It is desirable to reduce flame light by the introduction of a cyan filter between the flame and the camera. Photography in normal daylight conditions is again possible.

The sparks were focused onto the camera lens with a short focal length, 5 cm diameter achromatic condensing lens (as indicated in Fig. 8.19).

For the isothermal work, a Schneider Xenotar f3·5/135 mm lens was used, fitted into an MPP technical camera. The camera was focused onto a plane in the centre of the spray with a depth of field of 0·7 mm. This depth of field was checked by photographing a graticule used for calibrating microscopes.

When the flame was photographed it was found necessary to increase the camera lens objective distance to avoid heating effects. This rearrangement led to a reassessment of the criteria governing the choice of camera lens, a difficulty made more acute by a requirement that the depth of field be reduced simultaneously to avoid too many 'in focus' drops appearing on the plates. The relationship governing the choice of camera lens is

$$D = \frac{2(M + 1)C\lambda}{M^2} \qquad (8.51)$$

where

$D =$ depth of field of lens; $M = v/u =$ magnification; $C =$ circle of confusion (focal length/1000); $\lambda = f$ number of lens.

This expression serves to illustrate the problems of incompatibility involved in the choice of lens. It is noteworthy that the lens chosen must also be capable of resolving very small drops.

Present day developments are still in some state of flux, but seem to indicate the use of a Schneider Xenar f4·5/180 mm lens creating a depth of field of about 0·9 mm. This choice is dictated to some extent by the droplet density in the spray to be photographed.

The magnification on the plates was of the order of two for both isothermal and flame photography. For measurement purposes this was increased to 100 times by projection onto a translucent screen. This type of screen is ideal for the work, since it allows measurement on the rear side of the screen, thus removing the difficulty of blocking off the projected light. No loss of contrast in the image was observed in the back projection.

THE SPARK MONITORING UNIT

A knowledge of the time interval between the sparks is required for the calculation of velocity. The 'optical probe unit' made for this purpose consisted simply of a vacuum photo-emissive cell which was directly coupled with a 50 KΩ load resistor to a cathode follower valve (6AM6). The configuration was such that increasing current through the cell produced an equivalent positive shift of the valve grid and hence the cathode. The photocell and valve were enclosed in a brass tube provided with a narrow slit for light entry and connected by five conductors—earth, + 200 V, cathode potential (centre of coax. cable) and two 6·3 V heater lines as a twisted pair—to a small transformer HT unit. The probe was positioned to give it reasonable vantage of the flashes and the resulting wave form was observed or recorded on the monitor oscilloscope and polaroid camera. Although the probe was successful in its prime function, *i.e.* the identification of the flash instant, the waveshape of the decay of the light pulse was liable to be inaccurate as a result of grid current being drawn by the valve during positive drive.

As an additional diagnostic device, detector coils consisting of perspex bobbins ½ in diameter wound with 6–10 turns of single core wire were placed over each of the two earthy discharge capacitor leads. Correlation between these two systems for determining the time interval between sparks was very satisfactory.

THE ELECTRONICS SYSTEM

In order to provide adequately rapid control of the picture recording sequence, events must be controlled electronically.

A block diagram of the electronic system is shown in Fig. 8.21. The method used has been devised to give complete control of the system from the camera synchronising contact. The shutter speed is set to half a second,

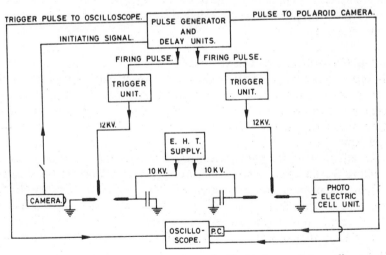

Fig. 8.21. *Block diagram of electronics (double image photograph).*

during which time the whole sequence of events takes place. The pulse generator and delay unit control the sequence of events—opening the shutter on the polaroid camera, triggering the oscilloscope and firing the spark units. The firing pulse closes the gate on a thyristor which allows the discharge of a capacitor into a pulse transformer, thus providing the triggering spark (12 kV) which ionises the atmosphere between the electrodes. The ionised atmosphere allows the almost instantaneous discharge of a capacitor charged to 10 kV. The light emission from each spark is picked up on the photo-electric cell unit, monitored on the oscilloscope and recorded by the polaroid camera.

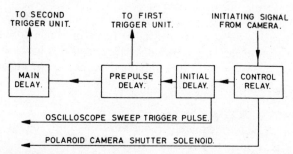

Fig. 8.22.   *Block diagram of pulse generator and delay unit* (*double image photograph*).

The oscilloscope used is a Tetronix 561A fitted with a 3A6 dual trace amplifier and 383 time base. The polaroid camera is a Tetronix C12RS. For two sparks it is possible to use a digital counter, but this is not practicable with more than two.

PULSE GENERATOR AND DELAY UNIT[29]

In order to provide the firing pulses to the spark gap control thyristors, it was necessary to construct an electronic unit which would provide all the signals from either the depression of a manual push button, or alternatively, from the shutter synchronising contact on the main camera. The circuit is illustrated in Fig. 8.22.

In addition to the pulses necessary for firing the thyristor gates controlling the flashes, a contact closure was required before the flashes to actuate the camera shutter solenoid on the monitor oscilloscope. Following this, a firing pulse was sent to the external time base trigger of the oscilloscope to start the sweep just before firing the main flashes. The action from the control unit may therefore be tabulated in chronological order:

(1) Camera contact closes as a result of manual operation of camera shutter through cable release.

(2) Control relay closes.

(3) 'Make' contact on control relay actuates monitor oscilloscope camera shutter solenoid. Second 'make' contact on control relay (in parallel with manual push button) excites a monostable delay circuit of 200 μsec duration.

(4) Termination of the above monostable delay generates a positive trigger pulse to the time base trigger of the monitor oscilloscope and therefore starts the sweep. The same positive pulse is also used to excite a variable delay monostable circuit of 20 to 120 μsec delay.

(5) Termination of (4) produces a pulse which fires the first main spark and simultaneously excites a further variable delay monostable circuit (three switched ranges from 10 to 4000 μsec).

(6) Termination of the second delay produces a positive pulse which fires the second main spark.

(7) 500 μsec after the instant of closure of the control relay it reopens, thereby releasing the monitor oscilloscope camera shutter solenoid. This latter operation terminates one sequence, whereupon the system is ready for re-use.

It should be noted that, although the above system has been described in terms of providing two 'main' flashes, the nature of the circuit allows the subsequent addition of further delays with the minimum of effort, since the whole operation works sequentially and any positive outgoing pulse can be used as the exciter pulse of the subsequent stage. However, it is noteworthy that introduction of further flashes in this way would lead to complication of the photographic recording of the events owing to prefogging of the photographic plate which increases with the number of flashes registered.

Further details of the electronic system, including full circuitry, is available in reference 29.

8.8 TURBULENCE MEASUREMENTS IN FLAMES

Turbulence measurements in flames are required for the determination of exchange coefficients for heat, mass and momentum transfer. Local reaction rates are not only dependent upon local air/fuel ratios, but also on the degree of unmixedness, which is in turn dependent upon the local turbulence intensity. Turbulence levels also affect the rate of mixing and entrainment from surrounding gases as well as flame stabilisation. Measurements are required of instantaneous values of velocity, temperature and concentration of species together with correlations of these quantities.

HIGH TEMPERATURE HOT WIRE PROBES

Two high temperature hot wire probes suitable for measurements in flames are available. One is the DISA 55A74 probe in which the sensor is an 80/20 platinum–iridium wire 10 in diameter and 1·5 mm long. This probe can be operated with a mean wire temperature as high as 940°C. Thermo-Systems Inc. produce a cooled probe which can be used in an environment up to 3000°C. The sensor of this probe is a glass tube 0·006 in od and 0·004 in id cooled with an internal flow of water or an inert gas. The resistance of a thin film of platinum on the outside surface of the tube is maintained constant by the associated circuitry. The major limitations of these probes is that they are not available with multiple sensors. Rao and Brzustowski[30] have shown recently that, with a single sensor probe, measurements can be made in a diffusion flame. The power dissipated in the sensor is given by

$$P = (A + BV^n)(T_s - T_e) \tag{8.52}$$

where

$$A = k_f C \left(\frac{T_f}{T_e}\right)^{0.17} \tag{8.53}$$

$$B = k_f D \left(\frac{d_0}{v_f}\right)^n \left(\frac{T_f}{T_e}\right)^{0.17} \tag{8.54}$$

In the constant temperature hot wire anemometer, the sensor temperature T_s is maintained constant while both the velocity, V, and the temperature, T_e, of the gas vary. In addition, the values of A and B depend upon the temperature and composition of the gas. The evaluation of A and B can be made tractable by assuming or measuring local gas concentrations. For the plume conditions studied by Rao and Brzustowski,[30] it was assumed that the temperature of the plume decreased mainly as a result of entrainment of ambient air and, therefore, the composition of the gas at any point was taken to be a function only of its temperature $\bar{T}_e$. A and B were then evaluated at the film temperature $\frac{1}{2}(T_s + T_e)$, with k_f and v_f estimated by the standard methods of molecular transport theory. No attempt was made to incorporate the dependence of these quantities on temperature fluctuations.

With velocity, temperature and composition all varying, it is extremely difficult and time consuming to produce a wide ranging calibration of the hot wire. Rao and Brzustowski therefore accepted the Collis and Williams[31] correlations. Velocity and temperature effects can be separated when the sensor is operated perpendicular to the direction of the mean flow at a given location, at two different temperatures. The following two equations are then solved simultaneously

$$\bar{P}_1 = (A_1 + B_1 U^n)(T_{s1} - \bar{T}_e) \qquad (8.55)$$

$$\bar{P}_2 = (A_2 + B_2 U^n)(T_{s2} - \bar{T}_e) \qquad (8.56)$$

Measurement of the turbulent quantities $\overline{u^2}, \overline{t_e^2}, \overline{ut_e}$ requires the measurement of the mean square values of the fluctuating components of the bridge power at three different sensor temperatures (overheat ratios) with the sensor perpendicular to the mean flow direction. The turbulent diffusion terms $\overline{uv}$ and $\overline{vt_e}$ can be measured if the sensor is operated successively at an inclination φ to the mean flow at two different sensor temperatures, and then at an inclination φ to the mean flow at the same two sensor temperatures. The full equations and results of measurements in a plume are given in reference 30.

OTHER METHODS FOR TURBULENCE MEASUREMENTS

Turbulence measurements in flames can be made by measurement of a wide range of physical and chemical properties such as pressure, velocity, temperature, conductivity, radiation, viscosity, concentration of stable and unstable species, behaviour of particles in suspension, etc. In a review of measurements of flame turbulence Günther[32] lists five requirements:

1. It must be possible to make measurements at points or regions of small volume within the flame.
2. Probes should be extremely small in order to avoid disturbance of flow properties at the measuring point and upstream. Absence of probes is most desirable.
3. The device must be suitable for use in high temperature gases.
4. The method should be suitable for flow fields with pressure gradients such as occur in combustion chambers, behind flame holders or in swirling flow.
5. Results should depend on one variable only and not on a combination of two or more variables.

Table 8.1 gives a list of devices that have been used for measurement of turbulence in flames. It lists the principle of measurement, names of

TABLE 8.1

Measurement of Turbulence in Flames

Principle	Authors	Refs	Type of observed flame[a]	Fuel and burner	Measured and calculated values[b]
1. *Optical* Light scattering	Williams, Hottel and Gurnitz	34	P	Natural gas–air. Nozzle	L_E
	Grover, Falls and Scurlock	35, 36	P	Natural gas–air. 2-dim. burner	S_{tu}; co-ordination of flame front and their fluctuations
Total emission, in wide range of wavelengths	Lyon	37	D	C_2H_2, C_4H_{10}. Tubes and 2-dim. burner	$Tu(\varepsilon T^4)$ [c]
	Günther and Simon	38, 39 40		City gas. Nozzle	$Tu(\varepsilon T^4)$, L_E
Spectral emission	Günther	32	—	—	$Tu(\varepsilon T^4)$, L_E
2. *Pressure* Static[d]	Becker and Brown	41	D	Propane. Tube	$Tu(v)$
	Eickhoff	42	D	City gas. Tube	$Tu(v)$
Total (microphone)	Ebrahimi	43	D	H_2. Tube	$Tu(v)$
(hot wire in cooled stream)	Recknagel	44	D	City gas. Nozzle	$Tu(u)$, $F(u)$ [e]
3. *Conductivity (special hot wire)*	Raper	45	D	CO. Tube	$Tu(c)$

	Method	Author	Ref.	System	P/D	Quantity
4.	*Diffusion of tracer*	Westenberg and Rice	46, 47	C_5H_{12}–air. Flame in enclosure	P	$Tu(v)$
		Prudnikov	48	C_6H_6–air. Partly enclosed flame with flame holder	P, D	$Tu(v)$, $\varepsilon_i L_L$
		Hersch	49	H_2O_2 Combustor	P	$Tu(v)$ f
5.	*Ionisation*	Karlovitz	50	Natural gas–air. Tube	P	S_{tu}, $Tu(u)$, $F(T)$
		Shchetinkov	51		P	L_E
		Vlasov	52		P	L_E, $F(T)$
6.	*Thermocouple with compensation*	Kunugi and Jinno	53	City gas. Tube	D	T'; $F(T)$
7.	*Calculation from time mean values*	Hawthorne, Weddel and Hottel	54		D	$Tu(c)$
		Richardson, Howard and Smith	55		D	$Tu(c)$
		Günther	33		D	ε_c, ε_i

a P = premixed; D = diffusion.
b L_E, L_L = turbulence scales (Eulerian, Lagrangian); S_{tu} = turbulence intensity; ε_c, ε_i = exchange coefficients (matter, momentum); $F(T)$ = spectrum of frequencies.
c Measurements across whole jet only.
d To be used in flow without large differences in static pressure.
e Behaviour of boundary layer near nozzle.
f Across thin layer of flame gases.

investigators and references. The type of flame (whether premixed or diffusion), the type of fuel and burner, together with the type of measurement made, are also listed in the table.

Most methods listed are suitable for observation of turbulence intensity as well as frequency distribution. Many devices are suitable for both diffusion and premixed flames. Schlieren-optics, interferometry, laser interferometry and other laser devices are also being examined as means of measuring turbulence in flames. Very few of the systems used satisfy all the five requirements listed above and a complete description of the turbulence in a system will require a combination of systems.

REFERENCES

1. Gosman, A. D., Pun, W. M., Runchal, A. K., Spalding, D. B. and Wolfshtein, M., *Heat and Mass Transfer in Recirculating Flows*, Academic Press, London, 1969. Also, Heat Transfer Section Research Reports, Dept. Mechanical Engineering, Imperial College, London.
2. Fristrom, R. M. and Westenberg, A. A., *Flame Structure*, McGraw-Hill, New York, 1965.
3. Gaydon, A. G. and Wolfhard, H. G., *Flames*, Chapman and Hall, London, 1960.
4. Weinberg, F. J., *Optics of Flames*, Butterworths, London, 1963.
5. Lawton, J. and Weinberg, F., *Electrical Aspects of Combustion*, Clarendon Press, Oxford, 1969.
6. Thring, M. W., *The Science of Flames and Furnaces*, Chapman and Hall, London, 1962.
7. Thring, M. W. and Beér, J. M., Symposium on Temperature Measurement, Inst. Mech. Engrs, London, April 1962.
8. Flames and Industry: First Symposium, 1957; Second Symposium, 1962; Third Symposium, 1966. Institute of Fuel, London.
9. Kissel, R. R., Int. Flame Res. Found., Ijmuiden, Holland, Tb-F72/9/4, 1959.
10. Beér, J. M., Chigier, N. A., Koopmans, G. and Lee, K. B., Int. Flame Res. Found., Ijmuiden, Holland, F72/2/9, 1965.
11. Chedaille, J., and Braud, Y., *Industrial Flames*, Vol. 1, Intn. Flame Res. Found., Arnolds, London, 1971.
12. Vizioz, J. P. and Leuckel, W., Int. Flame Res. Found., Ijmuiden, Holland, K20/a/43, 1969.
13. Lee, J. C. and Ash, J. E., *Trans. ASME* 1956, **78**, p. 603.
14. Macpharlane, J., NGTE, Pyestock, Farnborough, England. Private communication.
15. Davies, T. W., Ph.D. Thesis, Sheffield University, 1969.
16. Davies, P. O. A. L., 'Recent developments in hot wire anemometry', Int. Seminar on Heat and Mass Transfer, Hercey Novi, Yugoslavia, 1969.

17. Hinze, J. O., *Turbulence*, McGraw-Hill, New York, 1959.
18. Bryer, D. W., Walshe, D. E. and Garner, H. C., Aeronautical Research Council, R & M. No. 3037, Her Majesty's Stationery Office, London, 1958.
19. Ricou, F. P. and Curtet, R., Université de Grenoble, France (private communication).
20. Miller, R. D. and Comings, E. W., *J. Fluid Mech.* 1960, **7**(2), p. 237.
21. D. Bradley *et al.*, 'Measurements of high gas temperatures with fine wire thermocouples', *J. Mech. Eng. Sci.* 1968, **10**, p. 299.
22. Dixon-Lewis, G. and Isles, G. L., 'Flame structure and reaction kinetics III: measurement of temperature profiles in flames at atmospheric pressure', *Proc. Roy. Soc.* 1968, **A308**, p. 517.
23. Smith, S. R. and Gordon, A. S., *J. Phys. Chem.*, 1956, **60**, p. 759.
24. Goodridge, M., Jackson, R. and Thurlow, G. G., *Trans. Soc. Inst. Techn.* 1956, **8**, p. 3.
25. Beér, J. M. and Lee, K. B., *Öl und Gasfeuerung* 1964, **9**(2), pp. 96–108.
26. Beér, J. M. and Claus, J., *J. Inst. Fuel* 1962, **35**, p. 437.
27. Penny and Giles Ltd, Mudeford, Christchurch, Hants., England.
28. Mellor, R., Chigier, N. A. and Beér, J. M., ASME Gas Turbine Conference, Brussels, 1970. Also, Mellor, R., Ph.D. Thesis, University of Sheffield, 1969.
29. Mellor, R., Poole, B. and Taylor, D., Conference on Kinematic Measurements on Rapidly Moving Objects, University of Bradford, Sept. 1968.
30. Rao, U. K. and Brzustowski, T. A., *Combustion Science and Technology* 1969, **1**, p. 171.
31. Collis, D. C. and Williams, M. J., *J. Fluid Mech.* 1959, **6**, p. 357.
32. Günther, R., *J. Inst. F.* 1970, **43**, pp. 187–92.
33. Günther, R., *Chem.-Ing.-Techn.* 1969, **41**, Heft 5 + 6, pp. 315–22.
34. Williams, G. C., Hottel, H. C. and Gurnitz, R. N., Twelfth Symposium on Combustion, pp. 1081–92, The Combustion Institute, Pittsburgh, 1969.
35. Grover, J. H., Falls, E. N. and Scurlock, A. C. Ninth Symposium on Combustion, p. 21, Academic Press, London, 1963.
36. Scurlock, A. C. and Grover, H. H., Selected Combustion Problems, AGARD Comb. Coll., Cambridge, 1953, p. 215.
37. Lyon, J. B., Thesis, University of Delaware, 1953.
38. Günther, R. and Simon, H.-D., Twelfth Symposium on Combustion, pp. 1069–79, The Combustion Institute, Pittsburgh, 1969.
39. Simon, H.-D., *Chem.-Ing.-Techn.* 1968, **40**(1/2), pp. 65–71.
40. Simon, H.-D., *Chem.-Ing.-Techn.* 1968, **40**(3), pp. 121–8.
41. Becker, H. A. and Brown, A. P. G., Twelfth Symposium on Combustion, pp. 1059–68, The Combustion Institute, Pittsburgh, 1969.
42. Eickhoff, H., *Chem.-Ing.-Techn.* 1968, **40**(22), pp. 995–8.
43. Ebrahimi, I., *Chem.-Ing.-Techn.* 1968, **40**(15), pp. 769–71.
44. Recknagel, J., Diss. Univ. Karlsruhe (TH), 1969.
45. Raper, A. G., Thesis, Sheffield University, 1957.
46. Westenberg, A. A., *J. Chem. Phys.* 1954, **22**, p. 814.
47. Westenberg, A. A. and Rice, J. L., *Combustion and Flame* 1959, **3**(4), p. 459.
48. Prudnikov, A. G., Seventh Symposium on Combustion, pp. 575–82, Butterworths, London, 1959.
49. Hersch, M., *ARS J.* 1961, **31**, p. 39.

50. Karlovitz, B., Denniston, D. W., Knapschäfer, D. H. and Wells, F. E., Fourth Symposium on Combustion, p. 613, Williams and Wilkins, Baltimore, 1953.

51. Shchetinkov, E. S., Combustion in Turbulent Flow, Proc. Moscow Sem. on Comb., 1959, Israel Progr. Sc. Transl. Ltd, 1964.

52. Vlasov, K. P., Combustion in Turbulent Flow, Proc. Moscow Sem. on Comb., 1959, Israel Progr. Sc. Transl. Ltd, 1964.

53. Kunugi, M. and Jinno, H., Seventh Symposium on Combustion, pp. 942–8, Butterworths, London, 1959.

54. Hawthorne, W. R., Weddel, D. S. and Hottel, H. C., Third Symposium on Combustion, pp. 266–88, Williams and Wilkins, Baltimore, 1949.

55. Richardson, J. M., Howard, H. C. and Smith, R. W., Fourth Symposium on Combustion, pp. 814–8, Williams and Wilkins, Baltimore, 1953.

56. DISA, Operating Manual, Type D5501 Anemometer.

Index

Air blast rotary cup atomiser,
151–153
Angle (flame), 42
Angles of flight, droplet, 160
Angular momentum, 95, 106, 109, 111
Annular jets, 24
Apparent origin of jet, 13
Archimedes number, 208
Atomisation of liquid fuels, 151
Atomisers, 151, 152
Axial
    momentum, 95, 109, 111
    thrust, 106
    velocity, 79, 118, 119

Basic differential equations, 87
Blockage ratio, 73–75
Blow-off, 38, 55–57, 82
Bluff body flame stabilisation, 68, 82
Boundary layer equations, 92–93
Breakpoint length, 37, 38
Buoyant jets, 208
Burning
    constant, 182
    rate of droplets, 165, 166, 171
    velocity, 55

Categories of similarity, 197–198
Chemical kinetics, 170
Circulation, 105
Co-axial jets, 24
Co-flowing parallel stream, 17
Combustion in recirculation zones, 79
Concentration
    decay, 20, 22, 42

Concentration—*continued*
    distribution
        axial, 11
        radial, 12, 41
    measurement, 244
Confined
    flame length, 43
    jets, 26
Craya–Curtet
    similitude parameter, 33
    theory, 36
Critical
    boundary velocity gradient, 57
    Reynolds number, 39
Cylindrical
    bluff body, 70
    polar co-ordinate system, 90

Degree of swirl, 96
Dimensional analysis, 198
Dimensionless groups, 199–200
Dissipation function, 91
Divergent
    flame propagation, 57
    nozzles, 123–126
Double
    concentric jets, 207
    image photography, 248
Drag coefficient, 176
Drop
    formation in sprays, 158
    size measurement, 155–156, 247
    trajectories, 159
Droplet arrays, 179
Ducted axisymmetric jet, 27

261

Transverse jets, 23
Turbulence
  intensity, 63, 66, 67, 76, 78, 176
  measurement in flames, 253–258
  models, 87, 90
Turbulent
  burning velocity, 60–64, 67
  combustion, 60
  exchange coefficient, 87
  flame theories, 63–65, 67
  flux, 87
  free jet, 15
  jet diffusion flame, 40
Twin-fluid atomiser, 151, 184

V-shaped flame holders, 79–81

Velocity
  decay, 19, 22, 122
  distribution, 11
  profiles, 13
Venturi pneumatic pyrometer, 235,
    236
Volumetric reaction rate, 54
Vortex eye, 74
Vorticity, 104

Wake, 81
  bluff body, of, 73
Weak swirl, 117, 120, 121
Wrinkled flame theory, 63–65, 67–68

Zero streamline, 74